ANATOMY OF MOVEMENT

Anatomy of Movement

Blandine Calais-Germain

EASTLAND PRESS ◆ SEATTLE

Originally published as *Anatomie pour le mouvement,* Editions Desiris (France), 1985. Revised in 1991.

Library of Congress Cataloging-in-Publication Data

Calais-Germain, Blandine.
    [Anatomie pour le mouvement. English]
    Anatomy of movement / Blandine Calais-Germain; [English language edition translated by Nicole Commarmond]. —English language ed.
    p. cm.
    Translation of: Anatomie pour le mouvement.
    Includes index.
    LCCN: 93-71669
    ISBN: 0-939616-17-3
    1. Human mechanics—Popular works. 2. Musculoskeletal system—Anatomy—Popular works. 3. Dancing—Physiological aspects. 4. Exercise—Physiological aspects. 5. Sports—Physiological aspects. 6. Physical education and training—Physiological aspects   I. Title.

QP301.C3513 1993   612.76

Printed in the United States of America

13th Printing

English language edition edited by Stephen Anderson, Ph.D.

Book design by Gary Niemeier

Dedicated to Marie, Patrick, Jacques, Francois

Table of Contents

....

Foreword

Anatomists, for many centuries, were concerned almost exclusively with precise description of the body's structures. Inevitably, they began by treating the locomotor system in the same way as the internal organs, i.e., actual functions were either unknown or described independently of structure.

Gradually, around the beginning of the 20th century, anatomists began paying more attention to the actions of muscles and joints. Such functional studies remained at an elementary level for several decades. More recently, some researchers began looking at biomechanical properties (such as elasticity and resistance) of the locomotor system. However, these studies were focused on isolated components in the laboratory, not on how muscles and joints are used in "real life." Functional aspects were often viewed in terms of "efficiency," i.e., how to make the body an obedient instrument of various physical disciplines.

In physiotherapy, body movements are analyzed in terms of both neurophysiological and mechanical components, thus allowing better definition of therapeutic effects and the real mechanisms of movement.

Many people interested in physical disciplines such as dance, mime, theater, yoga, relaxation, etc. have come to physiotherapy looking for quantitative as well as qualitative analytical studies which would facilitate their practice. In this way, Blandine Calais-Germain began by studying dance and ended up studying physiotherapy.

The complementary nature of these two ways of dealing with human body movements is obvious. Blandine quickly realized that dancers could benefit greatly from a better understanding of their "inner" bodies. She devised a novel teaching method to serve this purpose: the simultaneous representation of physical structures and their possible movements, designed to facilitate actual execution by the dancer.

Not only dancers, but also professionals involved in other physical disciplines, came in increasing numbers to her classes. The emphasis in these classes (and this book) is on anatomy not for its own sake, but for better understanding of body movements.

I have taken great pleasure in witnessing the birth of this concept, the first classes, and now the publication of this book which embodies Blandine's many years of experience as a dancer and teacher. I am delighted that the fruits of this experience will now be made widely available to others. Having worked closely with Blandine when she was a student of physiotherapy, I can attest to her skills as a therapist, her intelligence, and her love for teaching.

The drawings in this book are all original, and the emphasis is always on description and understanding of natural postures and movements. The book will be particularly useful to those who, by profession, deal with integrated or complex movements of the body. For those who deal with human anatomy in any way, it will provide a useful and thought-provoking resource. I wish for this book the great success it deserves.

Dr. Jacques Samuel
Director, French School of Orthopedics and Massage
Paris, France

Preface

....

I would like to briefly describe the content and organization of this book.

This is intended simply as an introductory text. The emphasis is on basic human anatomy as it relates to *external body movement*. Therefore, we will be concerned mainly with bones, muscles, and joints. There will be no description of the skull, visceral organs, circulatory system, central nervous system, etc.

The book is designed to be as compact as possible, and to avoid repetition. Thus, format may vary from one chapter to the next. Parts of the body that are affected by the same muscles may be described together. Reference may be given to a previous page where a certain structure or function is described in more detail.

For consistency and ease of orientation, drawings usually show structures from the right side of the body. Exceptions are clearly indicated.

Joints are sometimes drawn without the adjacent bones, so that the articular surfaces can be more clearly seen. Similarly, each muscle is drawn in isolation (without surrounding muscles) to make its function more obvious.

Chapter 1 provides basic orientation and terminology, and should be read first. Subsequent chapters are arranged in a logical order (starting with the trunk, moving down the arm, and then down the leg), and I recommend that they be read in this order. However, the reader with previous knowledge of anatomy may start at any chapter.

The index will be helpful for locating the page where a particular structure is first mentioned, or described in detail.

CHAPTER ONE

Introduction

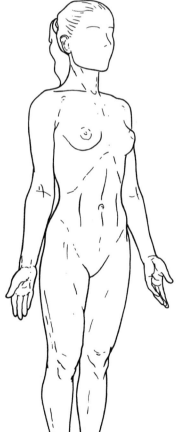

Anatomical position

The anatomy of movement, in the human body, involves interaction of three systems:

- the **bones,**
- linked together at the **joints,**
- are moved by action of the **muscles.**

Description of movements can be difficult. Various parts of the body can move in many different directions. Often more than one joint is involved. For consistency, the following conventions are generally followed:

- we begin by considering each joint in isolation;
- three perpendicular planes are used for reference;
- movements are described in relation to a standard "anatomical position" in which the body is standing upright, the feet parallel, the arms hanging by the sides, and the palms and face directed forward (see illustration).

Example: flexion of the wrist is a movement that takes the hand forward from the anatomical position.

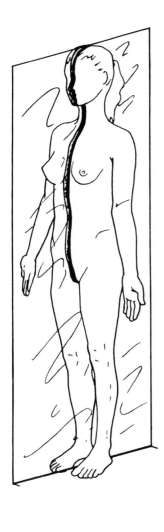

Planes of movement

The median, or midsagittal plane divides the body into symmetrical right and left halves.

Any plane parallel to the median plane is called a sagittal plane.

A movement in a sagittal plane which takes a part of the body forward from anatomical position is called **flexion**.

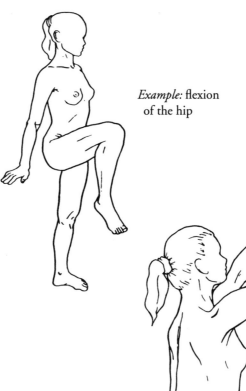

Example: flexion of the hip

Example: flexion of the shoulder

Exception: flexion of the knee

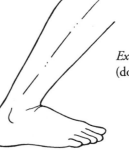

Exception: flexion (dorsiflexion) of the ankle

A movement in a sagittal plane which takes a part of the body backward from anatomical position is called **extension**.

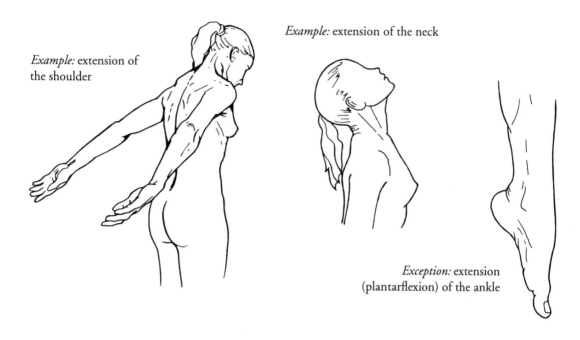

Example: extension of the shoulder

Example: extension of the neck

Exception: extension (plantarflexion) of the ankle

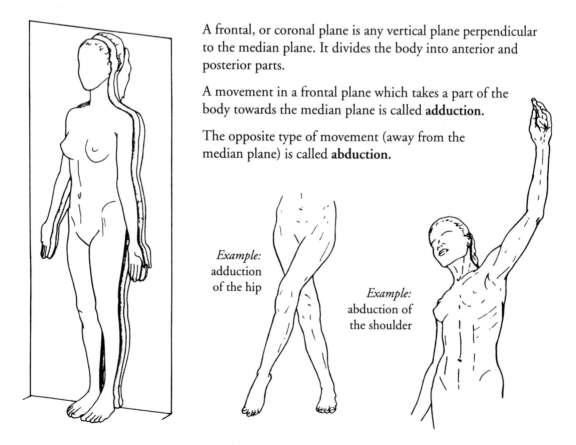

A frontal, or coronal plane is any vertical plane perpendicular to the median plane. It divides the body into anterior and posterior parts.

A movement in a frontal plane which takes a part of the body towards the median plane is called **adduction**.

The opposite type of movement (away from the median plane) is called **abduction**.

Example: adduction of the hip

Example: abduction of the shoulder

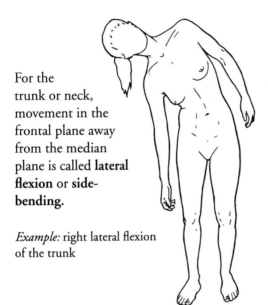

For the
trunk or neck,
movement in the
frontal plane away
from the median
plane is called **lateral
flexion** or **side-
bending.**

Example: right lateral flexion
of the trunk

For the fingers or toes,
the reference used is the
axis of the hand (middle
finger) or foot (2d toe).

Example: abduction of the
fifth finger moves it away
from the axis of the hand
(not from the median
plane).

A transverse, or horizontal plane divides the body into
superior and inferior (upper and lower) parts.

A movement in a transverse plane which takes a part
of the body outward is called **lateral rotation.**

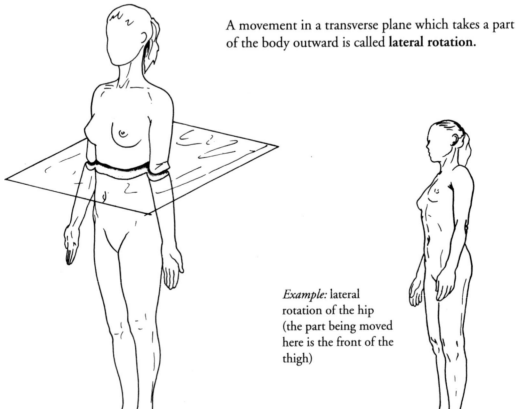

Example: lateral
rotation of the hip
(the part being moved
here is the front of the
thigh)

The opposite type of movement (inward) is called **medial rotation.**

Example: medial rotation of the shoulder (the part being moved is the front of the upper arm)

In **pronation** of the forearm, the palm of the hand faces backward.

In **supination** of the forearm, the palm faces forward.

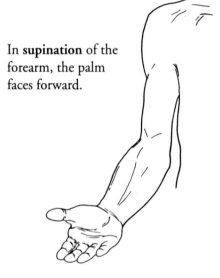

For the trunk or neck, we refer simply to right or left rotation. The reference point is the front of the chest or head.

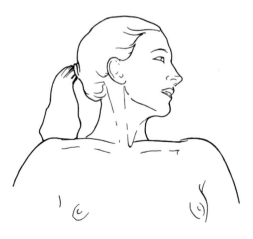

Complex body movements typically involve movement in all three planes.

Example: sitting in the "tailor's position" involves flexion, abduction, and lateral rotation of the hip joints.

Other anatomical reference terms

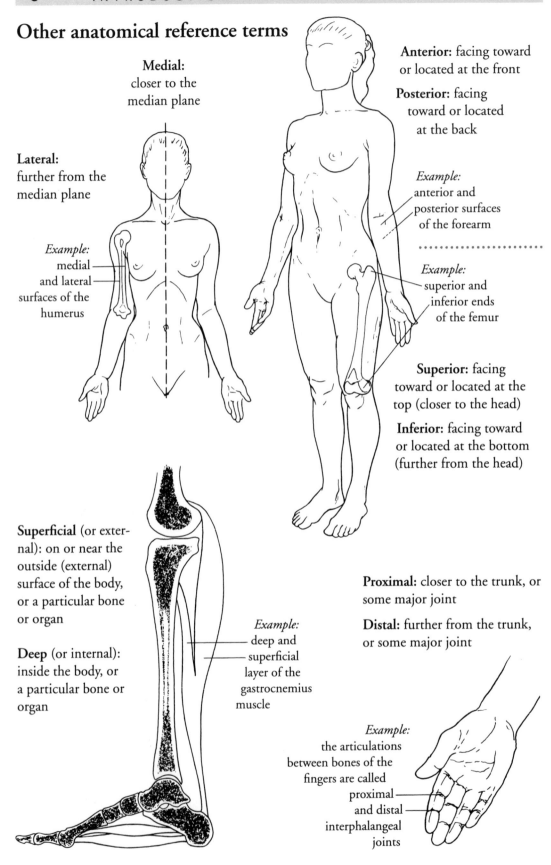

Medial: closer to the median plane

Anterior: facing toward or located at the front

Posterior: facing toward or located at the back

Example: anterior and posterior surfaces of the forearm

Lateral: further from the median plane

Example: medial and lateral surfaces of the humerus

Example: superior and inferior ends of the femur

Superior: facing toward or located at the top (closer to the head)

Inferior: facing toward or located at the bottom (further from the head)

Superficial (or external): on or near the outside (external) surface of the body, or a particular bone or organ

Deep (or internal): inside the body, or a particular bone or organ

Example: deep and superficial layer of the gastrocnemius muscle

Proximal: closer to the trunk, or some major joint

Distal: further from the trunk, or some major joint

Example: the articulations between bones of the fingers are called proximal and distal interphalangeal joints

Skeleton

The skeleton is a mobile framework of bones providing rigid support for the body.

The bones also serve as levers for the action of muscles.

There are three basic shapes of bones:

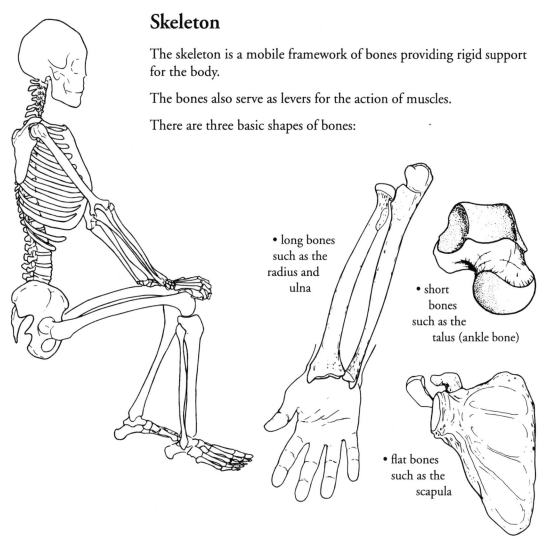

• long bones such as the radius and ulna

• short bones such as the talus (ankle bone)

• flat bones such as the scapula

Bone tissue consists of about ⅔ mineral components (mostly calcium salts) which give rigidity, and ⅓ organic components which give elasticity. Both these qualities are essential. Without rigidity bones would not keep their shape, but without some elasticity they would shatter too easily.

Bones are subjected to several types of mechanical strain:

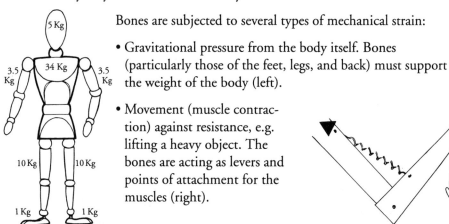

• Gravitational pressure from the body itself. Bones (particularly those of the feet, legs, and back) must support the weight of the body (left).

• Movement (muscle contraction) against resistance, e.g. lifting a heavy object. The bones are acting as levers and points of attachment for the muscles (right).

• Gravitational pressure from external objects (traction), e.g. supporting a heavy suitcase

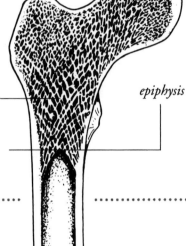

alveolar structure

epiphysis

Internal anatomy of a bone

Bones have evolved to withstand all these types of strain. We can see how by looking at a bone in cross section.

A long bone, in this case the femur, consists of three parts. They develop separately but are fused in the adult:

• the central shaft is called the **diaphysis**
• the two ends are called the **epiphyses**.

The diaphysis is a hollow tube with walls made of compact bone. The hollow structure gives light weight and is actually sturdier than a solid structure would be. Compact bone is thickest in the middle section of the diaphysis where mechanical strains are greatest.

A cross section of the epiphysis shows an alveolar (spongy) structure. Fibers are arranged in rows along the lines of greatest mechanical stress.

The hollow part of the diaphysis contains the bone marrow, where blood cells are manufactured. The marrow is red in children, but becomes yellow in adults as much of it is replaced by fatty tissue.

The external surface of the bone is covered with a membrane called the **periosteum,** which carries blood vessels and functions in bone repair.

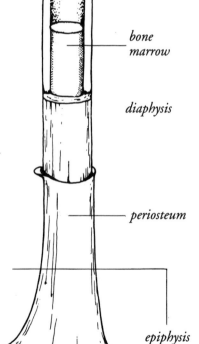

bone marrow

diaphysis

periosteum

epiphysis

Joints

Joints are areas where bones are linked together. They have varying degrees of mobility.

In some joints the bones are linked simply by fibrous connective tissue or cartilage. This allows little or no movement. These joints are not of great interest in a book about movement, but we will mention them occasionally.

Our primary interest will be in freely-movable joints (called diarthroses, or synovial joints) which contain a fluid-filled cavity between the articulating surfaces. These surfaces (sometimes called facets) are shaped so as to fit together but also allow movement. There are many general categories of joints, based on the shape of the articulating surfaces.

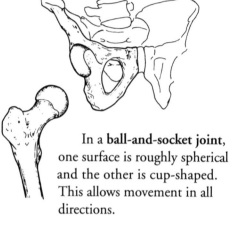

In a **ball-and-socket joint**, one surface is roughly spherical and the other is cup-shaped. This allows movement in all directions.

Examples: **hip**, shoulder

Hinge: the convex surface of one bone fits against the concave surface of the other in a clasping arrangement. Movement chiefly in one plane (flexion/extension).

Examples: elbow, knee, interphalangeal, metacarpalphalangeal, **ankle**

Gliding: both surfaces are essentially flat and movement is limited.

Examples: **intercarpal**, intertarsal, rib-vertebra, clavicle-scapula

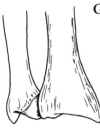

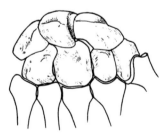

Ellipsoid: an oval-shaped process of one bone fits into a roughly elliptical cavity of the other. Movement in two planes (flexion/extension, abduction/adduction).

Examples: radius-carpals (wrist), **atlas-occipital**

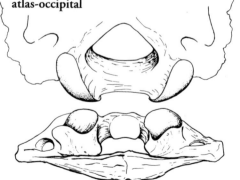

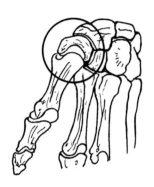

Pivot: a pointed or rounded process of one bone fits into a ring-like structure. Rotation is chief movement.

Example: atlas-axis

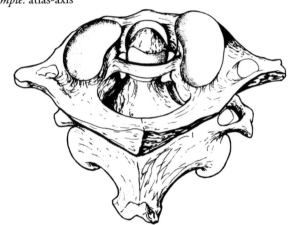

Saddle: both surfaces are saddle-shaped, i.e., convex in one direction and convex in the other. Movement in two planes.

Example: carpal-metacarpal I (thumb)

The articulating surfaces in a joint do not always make a snug fit. Some joints are more stable, and less likely to be injured, than others.

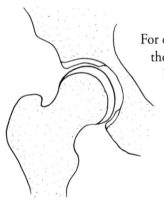

For example, the ball-and-socket structure of the hip is deep, snug-fitting, and protected by many strong muscles.

In contrast, the ball-and-socket of the shoulder is shallow, looser, less stable, and much more susceptible to injury.

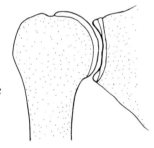

Between the articulating ends of the two bones in a joint is a gap which is not opaque to X-rays. This corresponds to the articular cartilages and synovial cavity.

In a dislocation or subluxation, a bone is moved from its normal position in a joint because of some trauma. There is associated damage to ligaments, etc. Dislocations are most common in finger, thumb, and shoulder joints.

Cartilage

Articulating surfaces of bones are covered with a
shiny, whitish connective tissue called cartilage.
It contributes to the synovial capsule and also
protects the underlying bone tissue.

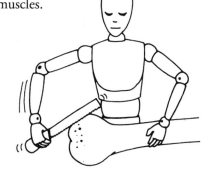

Example: cartilage
of the head of the
femur

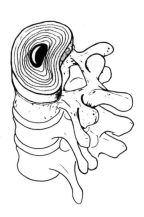

When movement occurs, joint cartilage
is subjected to two types of stress: gravita-
tional pressure (particularly in the weight-bearing joints of the legs and
feet), and friction from the movement itself. Cartilage is well-adapted to
these stresses, being strong, resilient, and smooth. Thus, it can absorb
shock and allow some sliding of the bones relative to each other.
Nevertheless, cartilage may be damaged either by trauma or excessive
wear (e.g., when the ends of the articulating bones do not provide a
good "fit"). Rheumatoid arthritis and osteoarthritis are two common
diseases involving damage to joint cartilage,
accompanied by inflammation, pain, and
stiffness of the joint and surrounding muscles.

Joint cartilage (like all cartilage) does
not contain blood vessels. It receives
nutrients from the synovial fluid,
and from blood vessels of the perichondrium and periosteum.

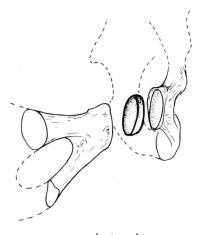

Fibrocartilage contains high concentrations of collagenous
(white) fibers and is specially adapted for absorbing shock.
It is found in the intervertebral discs, menisci (articular
discs) of the knee and other large joints, and symphysis pubis
(junction between the two pubic bones).

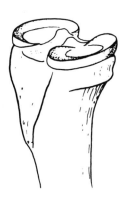

intervertebral discs   *menisci*   *symphysis pubis*

Joint capsule

This sleevelike structure encloses the joint, prevents loss of fluid, and binds together the ends of the articulating bones. The outer layer of the capsule is composed of dense connective tissue and represents a continuation of the periosteum. The inner layer, called the synovial membrane, is composed of loose connective tissue.

The joint capsule is stronger where movement must be prevented.

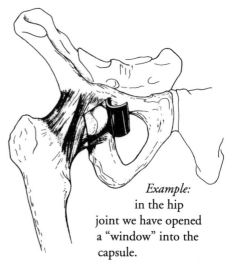

Example: in the hip joint we have opened a "window" into the capsule.

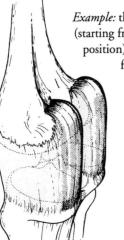

Example: the knee joint (starting from anatomical position) allows only flexion.

The capsule is strongly reinforced posteriorly to prevent extension. Fibers of the outer capsule are often arranged in parallel bundles (called **ligaments**; see following section) to reinforce joints and prevent unwanted movement.

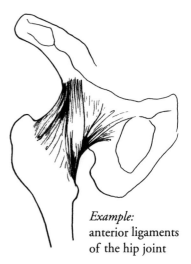

Example: anterior ligaments of the hip joint

The capsule may be arranged loosely or in folds where movement is allowed.

The **synovial membrane** secretes synovial fluid (shown by gray in the drawing), which fills the articular cavity. This fluid lubricates the joint, provides nutrients to the cartilage, and contains phagocytic cells which remove debris and microorganisms from the cavity. The actual amount of synovial fluid is surprisingly small: only about 3.5 ml (0.1 oz) in the knee, and much less for most joints.

Example: the capsule of the knee is loose in the front to allow flexion, and becomes folded during extension.

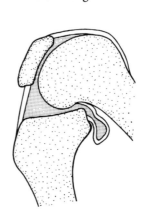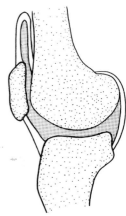

flexion   *extension*

Ligaments

Ligaments are dense bundles of parallel collagenous fibers. They are often derived from the outer layer of the joint capsule, but may also connect nearby but non-articulating bones.

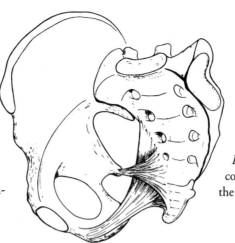

Example: the ligament connecting the sacrum to the spine of the ischium

The ligaments function chiefly to strengthen and stabilize the joint in a passive way. Unlike the muscles, they cannot actively contract. Nor (except for a few ligaments which contain a high proportion of yellow elastic fibers) can they stretch. They are placed under tension by certain positions of the joint and slackened by others.

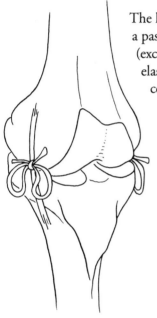

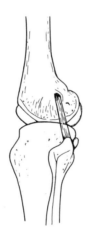

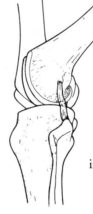

Example: the fibular collateral ligament of the knee is pulled tight in extension, and loosened in flexion.

extension   *flexion*

Ligaments contain numerous sensory nerve cells capable of responding to the speed, movement, and position of the joint, as well as to stretching or pain. The sensory cells constantly transmit such information to the brain, which in turn sends signals to the muscles via motor neurons. In this way, the body is usually able to avoid damage or undue stress to ligaments through appropriate corrective action. Nonetheless, excessive movement or trauma may result in spraining or rupture of ligaments.

Example: the most common type of knee injury in football is rupture of the tibial collateral ligament, often caused by a blow to the lateral side of the knee joint.

Muscle tissue

Essentially all movements of the human body result from contraction of muscles. In this book we are concerned with external movements, and will therefore focus on the **skeletal muscles** (also known as voluntary or striated muscles) which attach to bones. We will not discuss smooth muscle (which controls movements of the intestines, blood vessels, glands, etc.) or cardiac muscle (which causes the heart to beat).

In a microscopic cross section, we see that a muscle is composed of bundles of fibers (primary, secondary, tertiary), held together and compartmentalized by fibrous partitions called (on a progressively smaller scale) deep fascia, epimysium, perimysium, and endomysium. These connective tissue partitions (which are continuous with each other) allow easy movement of one muscle or muscle group relative to another. They can be extended beyond the muscle to form a strong fibrous cord called a **tendon** which is continuous with the periosteum of a nearby bone, and serves to attach the muscle to the bone. A broad, flattened tendon is called an **aponeurosis**.

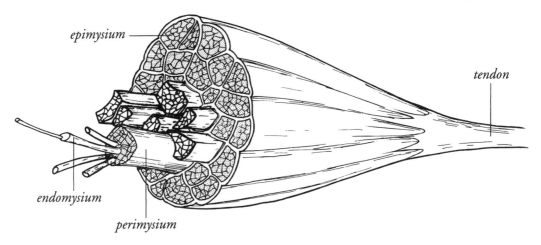

epimysium

tendon

endomysium

perimysium

Individual muscle cells (myofibers) are extremely long and, unlike most cells, contain many nuclei. Each cell contains many functional units called **sarcomeres**, divided by boundaries called Z lines (shown as vertical bars in picture). Each sarcomere contains **thick filaments** (made of the protein myosin) and **thin filaments** (protein actin); the thin filaments are anchored to the Z lines.

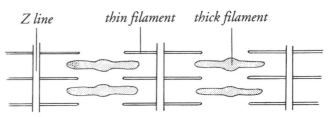

Z line   *thin filament*   *thick filament*

When the muscle is stimulated by a nerve, a series of chemical reactions involving calcium and ATP takes place, causing the thin filaments to "slide" along the thick filaments. As a result, the Z lines move closer together, and each individual sarcomere (and therefore the entire muscle) becomes shorter. This is the basis of muscle contraction.

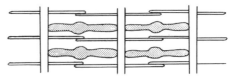

Typically, a muscle is attached to two different bones. For a given body movement, one bone (called the **origin**) is fixed in some way, and the other (called the **insertion**) moves as a result of muscle contraction. The origin is often the proximal bone, and the insertion the distal bone, but there are many exceptions.

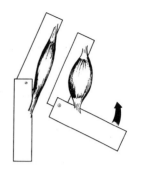

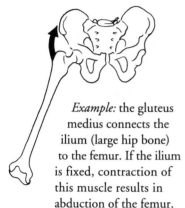

Example: the gluteus medius connects the ilium (large hip bone) to the femur. If the ilium is fixed, contraction of this muscle results in abduction of the femur.

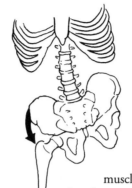

Example: on the other hand, if one is standing with the weight on the leg, the femur becomes the fixed point, and contraction of the muscle results in lateral flexion of the pelvis. In discussing muscle actions, we will generally assume that the proximal attachment is the origin, and provide additional description of a distal fixed point only when appropriate.

Besides their (active) ability to contract, muscles have a (passive) property of elasticity. When stretched, they tend to return to their normal resting length.

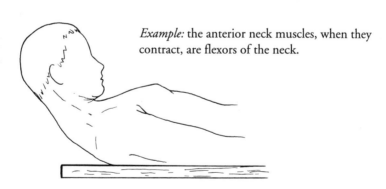

Example: the anterior neck muscles, when they contract, are flexors of the neck.

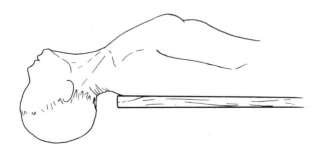

Example: during extension of the neck, they become stretched. When this happens, because of their elasticity, they tend to return the head to its anatomical position.

Muscle shapes

Muscles attach to bones in several manners:

- via an aponeurosis (broad tendon)

  Example: quadratus lumborum

- directly via muscle fibers (usually in a broad insertion)

  Example: subscapularis

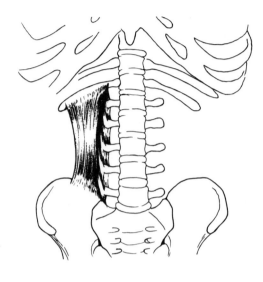

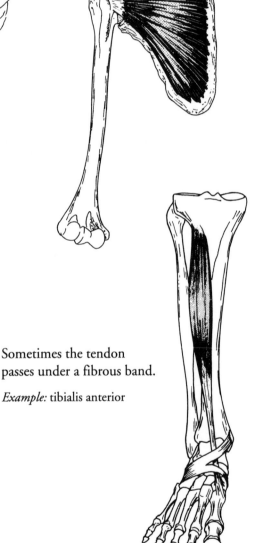

- via a regular tendon

  Example: brachioradialis

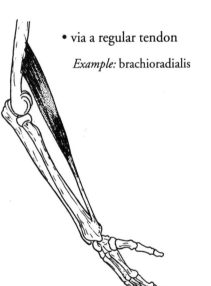

Sometimes the tendon passes under a fibrous band.

Example: tibialis anterior

Some muscles have several origins (called "heads"), which may be on more than one bone. For example, biceps brachii has two heads, triceps brachii has three heads, and quadriceps femoris has four heads. The flexor digitorum superficialis originates from both the radius and ulna. Multiple insertions are less common than multiple origins, and usually involve finger and toe bones.

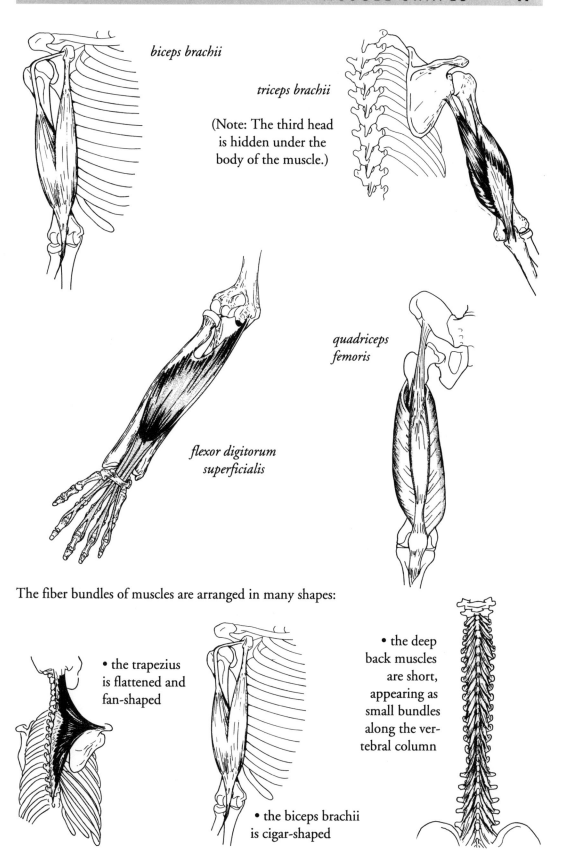

biceps brachii

triceps brachii

(Note: The third head is hidden under the body of the muscle.)

quadriceps femoris

flexor digitorum superficialis

The fiber bundles of muscles are arranged in many shapes:

• the trapezius is flattened and fan-shaped

• the deep back muscles are short, appearing as small bundles along the vertebral column

• the biceps brachii is cigar-shaped

Depending on the orientation and attachment of their fibers, muscles may act in one or several directions.

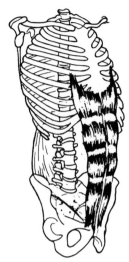

Example: the fibers of the rectus abdominis run essentially parallel to each other. This muscle flexes the trunk.

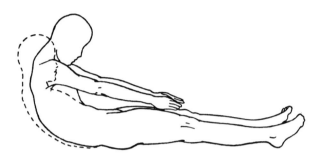

Example: the fibers of the external oblique (located bilaterally on the sides of the abdomen) are arranged like a fan. This muscle can produce anterior flexion, side-bending, or rotation of the trunk.

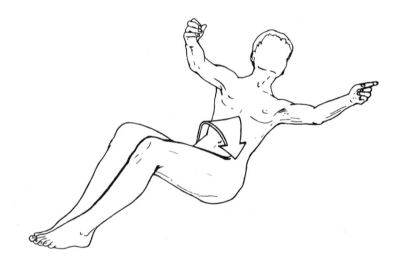

Long muscles (e.g., those inserting on the femur or tibia) are usually kinetic, i.e., able to produce highly visible external motion. Short, deep muscles (e.g., those inserting on the vertebrae or foot bones) tend to be responsible for precise, small-scale adjustments rather than gross movements.

A muscle which crosses and affects a single joint is called **monoarticular.** A muscle which crosses (and moves) more than one joint is called **polyarticular.**

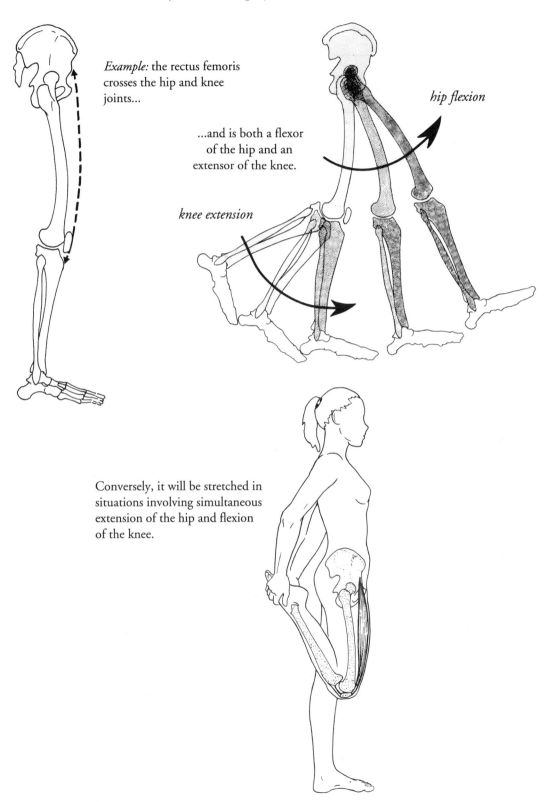

Example: the rectus femoris crosses the hip and knee joints...

...and is both a flexor of the hip and an extensor of the knee.

hip flexion

knee extension

Conversely, it will be stretched in situations involving simultaneous extension of the hip and flexion of the knee.

Muscle contraction

When we speak of a particular movement, the muscle which produces it is called an **agonist**, and a muscle which produces the opposite movement is called an **antagonist**.

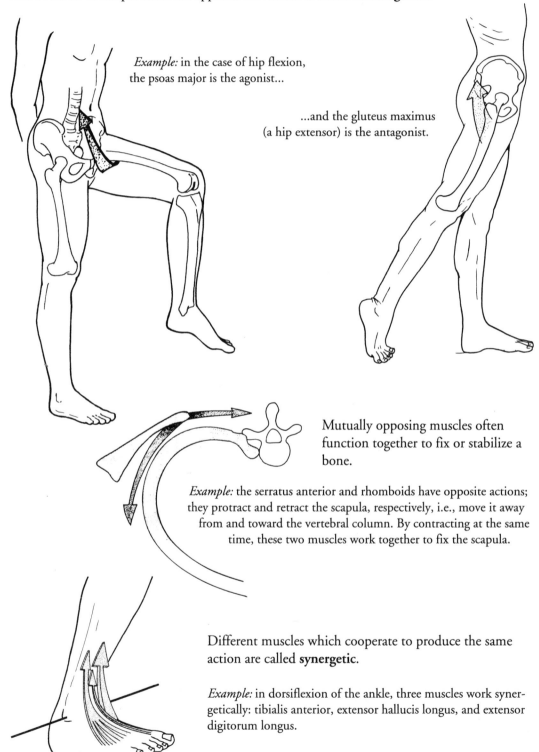

Example: in the case of hip flexion, the psoas major is the agonist...

...and the gluteus maximus (a hip extensor) is the antagonist.

Mutually opposing muscles often function together to fix or stabilize a bone.

Example: the serratus anterior and rhomboids have opposite actions; they protract and retract the scapula, respectively, i.e., move it away from and toward the vertebral column. By contracting at the same time, these two muscles work together to fix the scapula.

Different muscles which cooperate to produce the same action are called **synergetic**.

Example: in dorsiflexion of the ankle, three muscles work synergetically: tibialis anterior, extensor hallucis longus, and extensor digitorum longus.

When a muscle contracts, it tends to draw its origin and insertion points closer together. Anything that opposes this tendency is called **resistance**. For example, the brachialis and biceps brachii are the major flexors of the elbow. Their action can be opposed by several types of resistance:

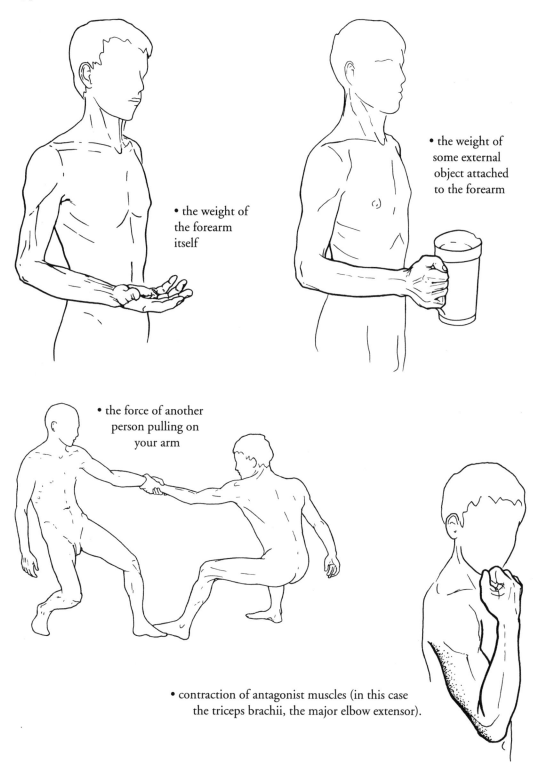

• the weight of the forearm itself

• the weight of some external object attached to the forearm

• the force of another person pulling on your arm

• contraction of antagonist muscles (in this case the triceps brachii, the major elbow extensor).

A given movement may be produced in different ways depending on the position of the body.

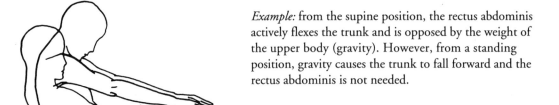

Example: from the supine position, the rectus abdominis actively flexes the trunk and is opposed by the weight of the upper body (gravity). However, from a standing position, gravity causes the trunk to fall forward and the rectus abdominis is not needed.

When a muscle actually shortens in length, and its origin and insertion are drawn together (e.g., in lifting a book), we use the term "**isotonic** contraction." This means that the force of contraction is constant and the length of the muscle changes. When a muscle strains against some resistance but does not change in length (e.g., in pushing outward against a doorframe with both hands), we use the term "**isometric** contraction," which means that the length stays the same but the force changes.

In the case of trunk flexion from a supine position, as shown above, the rectus abdominis is undergoing isotonic contraction.

Example: two men (A and B) are clasping hands and each is attempting to pull the other forward. As long as neither moves, the muscles of their elbow and shoulder joints are undergoing isometric contraction. As soon as one "wins" (in this case, A), his elbow flexors undergo isotonic contraction.

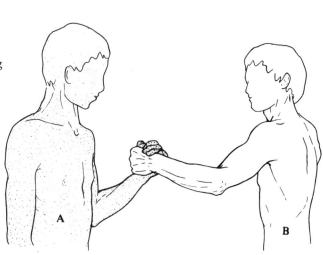

A

B

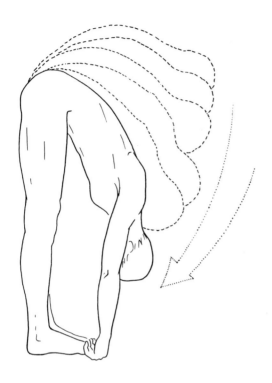

It is also possible for a muscle to be working even when it is increasing in length. This means that it is opposing or acting as a "brake" for the movement actually taking place, which would otherwise happen faster.

Consider again the example of trunk flexion from a standing position. Gravity is pulling the trunk forward. If all the muscles of the body were limp, this flexion would happen quickly. However, the movement can be slowed down by action of trunk extensors (e.g., quadratus lumborum). The extensors, even though they are being stretched, are actively attempting to contract.

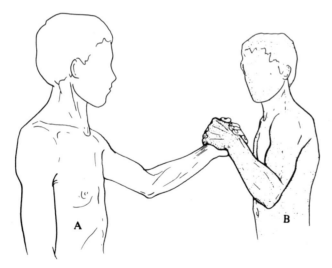

Or, in the tug-of-war example, A "loses" but action of his elbow flexors slows down or "brakes" B's elbow flexion.

During squatting or a "grand plie," flexion of both the hip and knee joints is slowed down by isometric contraction of the rectus femoris and hamstring muscles. Although the bones of the upper and lower leg are changing their position, and the muscles are active, the length of the muscles does not increase or decrease significantly. In this case, the simultaneous flexion of the hip and knee "cancels out" the change in muscle length which would occur if only one of these joints were involved.

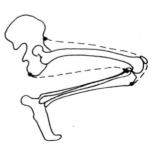

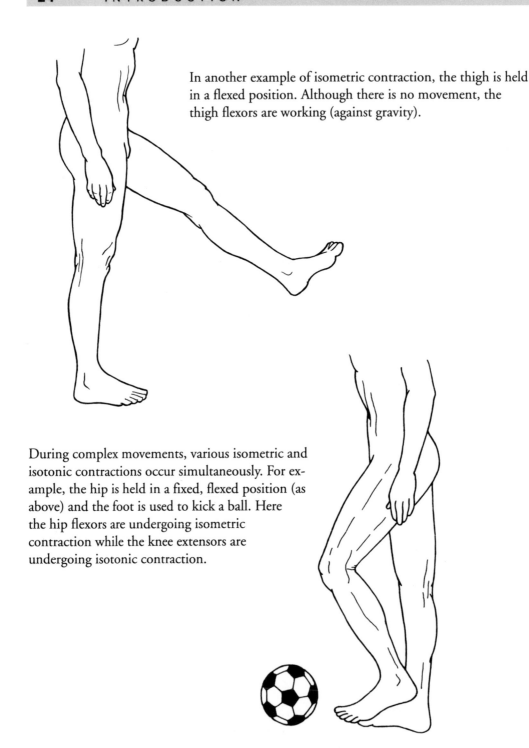

In another example of isometric contraction, the thigh is held in a flexed position. Although there is no movement, the thigh flexors are working (against gravity).

During complex movements, various isometric and isotonic contractions occur simultaneously. For example, the hip is held in a fixed, flexed position (as above) and the foot is used to kick a ball. Here the hip flexors are undergoing isometric contraction while the knee extensors are undergoing isotonic contraction.

CHAPTER TWO
The Trunk

....

The trunk is the central part of the body. In this book, we will examine only its locomotor functions, not its internal organs.

The **vertebral column**, or backbone, has important dual functions. On one hand, it can bend and rotate in many directions, thanks to its 23 intervertebral articulations. In this respect, its mobility can be compared to that of a measuring tape, in contrast to the limbs, whose angular movements resemble those of a folding measuring stick.

On the other hand, the vertebral column contains a central tunnel through which runs the delicate **spinal cord**. The nerve roots which supply the muscles and bring in sensory information branch off the spinal cord and exit through small openings between the vertebrae. Thus, the vertebral column has an important protective function, and it is essential that the vertebrae remain properly aligned and stabilized even during complex movements or stresses.

This dual function depends on a finely integrated system of mostly polyarticular muscles, which are either deep (composed of numerous small bundles) or superficial (usually arranged like broad sheets). There are also numerous ligaments holding vertebrae together.

Movements of the **pelvis** are difficult to separate from those of the vertebral column. In fact, the sacrum (the posterior part of the bony pelvis) consists of five expanded and fused vertebrae. The **ribs** articulate posteriorly with the vertebral column and anteriorly with the **sternum** or breastbone. Therefore, these bony structures will be included in this chapter.

Landmarks

Some visible and palpable landmarks of the trunk are shown below.

FRONT VIEW

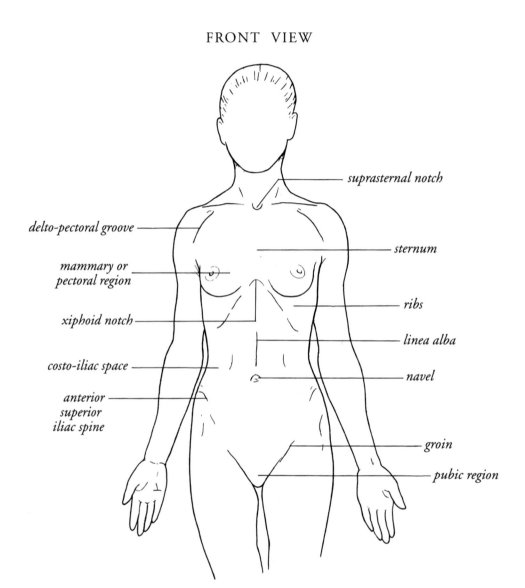

delto-pectoral groove

mammary or
pectoral region

xiphoid notch

costo-iliac space

anterior
superior
iliac spine

suprasternal notch

sternum

ribs

linea alba

navel

groin

pubic region

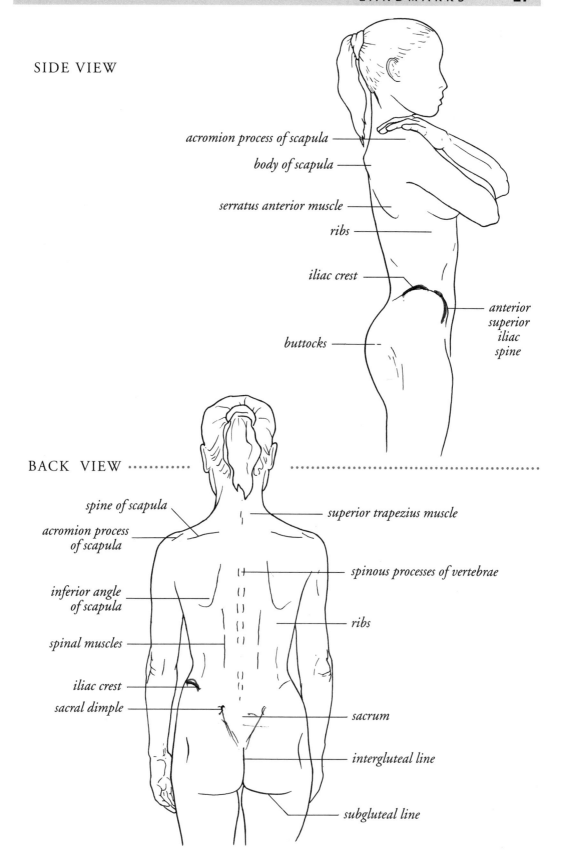

SIDE VIEW

acromion process of scapula

body of scapula

serratus anterior muscle

ribs

iliac crest

buttocks

anterior superior iliac spine

BACK VIEW

spine of scapula

acromion process of scapula

inferior angle of scapula

spinal muscles

iliac crest

sacral dimple

superior trapezius muscle

spinous processes of vertebrae

ribs

sacrum

intergluteal line

subgluteal line

Movements of the trunk

Thanks to the mobility of the vertebral column,
the trunk can move in several directions:

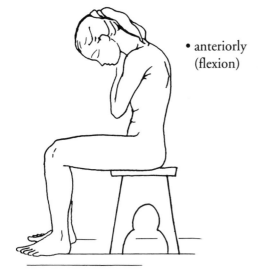

• anteriorly
(flexion)

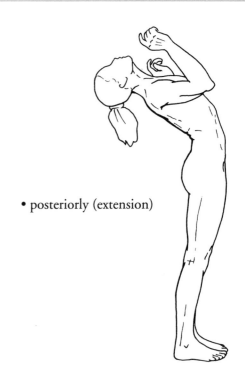

• posteriorly (extension)

• laterally (lateral flexion or
sidebending)

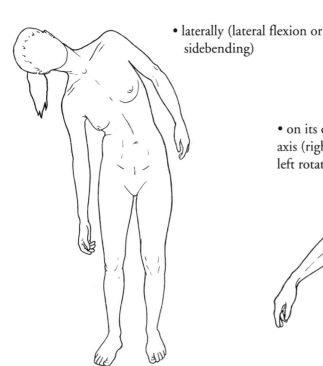

• on its own
axis (right or
left rotation)

Range of movement varies depending on vertebral level due to several factors:

• shape of the vertebrae (the lumbar vertebrae are the most massive and least mobile)
• thickness of intervertebral discs (the thicker the disc, the greater the mobility)
• the thoracic vertebrae articulate with ribs, which limit their mobility.

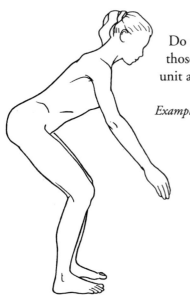

Do not confuse these movements with those in which the trunk moves as a single unit at the hip joint.

Example: hip flexion

Trunk movement can be a secondary consequence of limb movement.

Example: abduction of the arm takes the trunk into lateral flexion

The trunk provides the base for translation movements (called "isolations" in dance or mime, right).

These involve minimal sliding displacements of individual vertebrae, but the total resulting movement is large because of the many vertebrae involved (below).

Example: front-to-back or side-to-side movements of the head and pelvis while using a hula-hoop

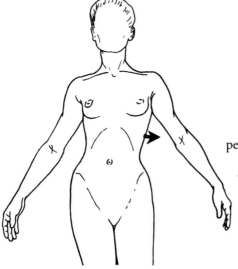

Often, two or more trunk movements are performed in combination.

Example: rotation, extension and lateral flexion

Vertebral column

We see obvious changes in size and shape of vertebrae as we examine the backbone from top to bottom:

- the seven **cervical** vertebrae are relatively small, and have holes (foramina) in their transverse processes

- the twelve **thoracic** vertebrae articulate with the twelve pairs of ribs

- the five **lumbar** vertebrae are massive, weight-bearing structures with limited mobility

- the **sacrum** consists of five fused, modified vertebrae, and articulates with the two ilium bones to complete the pelvic ring

- the **coccyx** or tailbone is a vestigial structure consisting of three or four fused vertebral remnants

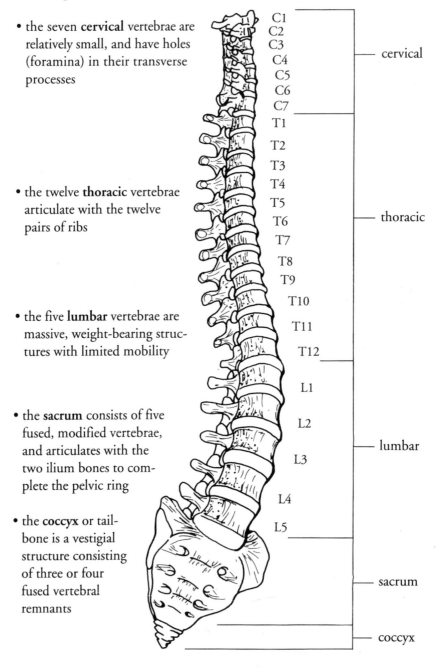

Within each region, vertebrae are numbered sequentially from top to bottom. For convenience, we usually refer to them by a letter plus a number. *Examples:* C7= seventh cervical vertebra; T3= third thoracic vertebra; L2= second lumbar vertebra; S1= first sacral vertebra, etc.

There are several characteristic **curvatures** of the vertebral column:

- sacrum, convex toward the back
- concave lumbar region (the term "lordosis" can refer either to an exaggeration of this curvature, or to the normal condition)
- convex thoracic region (kyphosis). An abnormal lateral curvature of the thoracic region is called scoliosis.
- concave cervical region.

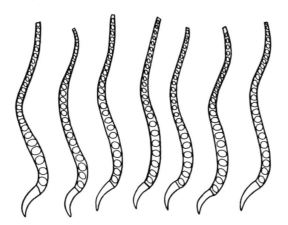

Exact form of these curvatures varies between people; this is normal. For example, kyphosis is almost non-existent in some individuals.

External appearance of these curvatures can be affected by overlying "soft" structures.

Example: a person with large buttocks may appear to have more pronounced lordosis than a person with small buttocks. However, X-rays could show identical lumbar curvatures.

Vertebral structure

Each vertebra consists of two main parts: the massive **body** (anterior), and the **vertebral arch** (posterior).

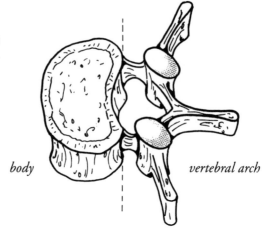

body   vertebral arch

The arch in turn can be divided into many parts. It is connected to the body by two **pedicles**. Two **lamina** unite posteriorly to form a **spinous process**. The thickened junctions between the pedicles and laminae have superior and inferior cartilaginous **articular facets** and a laterally-projecting **transverse process**.

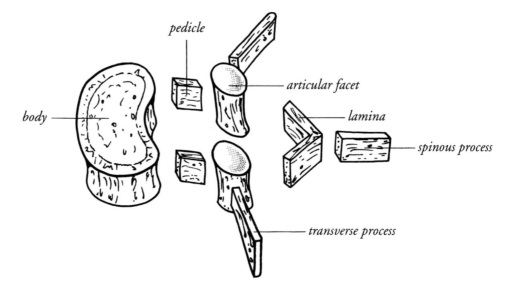

pedicle

body

articular facet

lamina

spinous process

transverse process

The opening between the body and the arch is called the vertebral foramen. As foramina of many vertebrae are lined up, they form the **vertebral canal** through which the spinal cord passes.

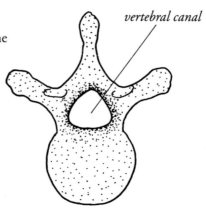

vertebral canal

The spaces between the pedicles of adjacent vertebrae form a series of openings called **intervertebral foramina.** As spinal nerves branch off the spinal cord, they exit through these foramina.

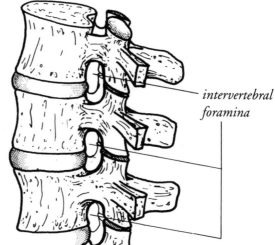

intervertebral foramina

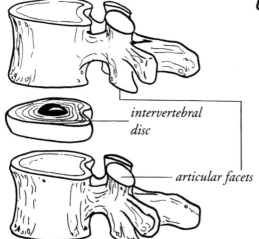

intervertebral disc

articular facets

Each vertebra is attached to its neighbor by three joints. The bodies are joined by the fibrocartilaginous **intervertebral disc.** Posteriorly, the two inferior articular facets of the top vertebra contact the two superior articular facets of the bottom vertebra. These facets are small, and serve mainly to guide movements.

nucleus pulposus

annulus fibrosus

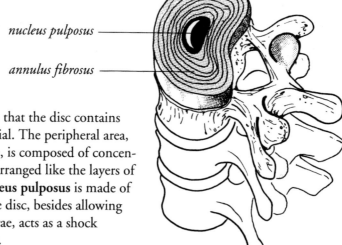

In cross section, we can see that the disc contains two distinct types of material. The peripheral area, called the **annulus fibrosus,** is composed of concentric rings of fibrocartilage arranged like the layers of an onion. The central **nucleus pulposus** is made of a gelatinous substance. The disc, besides allowing movement between vertebrae, acts as a shock absorber and weight bearer.

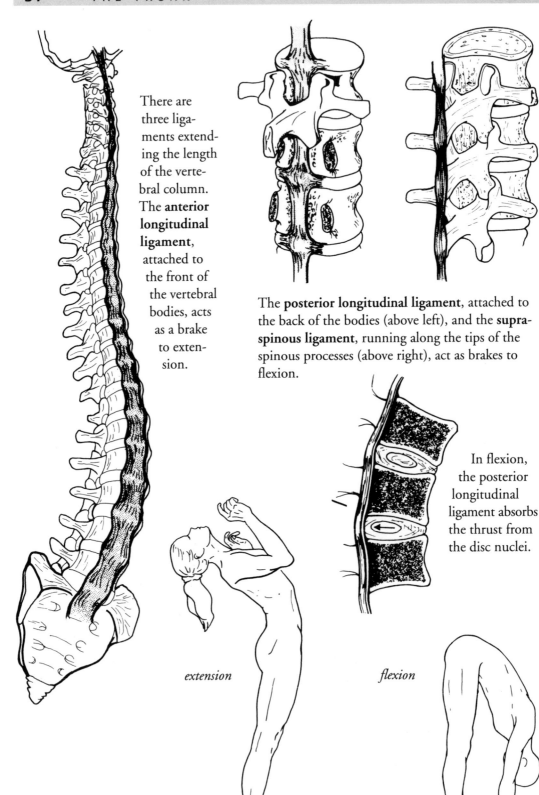

There are three ligaments extending the length of the vertebral column. The **anterior longitudinal ligament**, attached to the front of the vertebral bodies, acts as a brake to extension.

The **posterior longitudinal ligament**, attached to the back of the bodies (above left), and the **supraspinous ligament**, running along the tips of the spinous processes (above right), act as brakes to flexion.

In flexion, the posterior longitudinal ligament absorbs the thrust from the disc nuclei.

extension

flexion

Other vertebral ligaments are discontinuous. The **ligamenta flava** connect the laminae of adjacent vertebrae. These ligaments are elastic, and can be pierced by a syringe during a spinal tap.

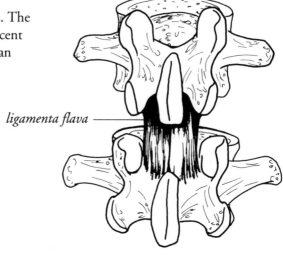

ligamenta flava —

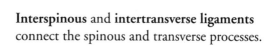

Interspinous and **intertransverse ligaments** connect the spinous and transverse processes.

intertransverse ligament

interspinous ligament

Right sidebending stretches the left intertransverse ligaments.

Other ligaments, specific to certain regions, will be mentioned later.

Vertebral movements

We can think of the vertebral column as a series of fixed segments (the vertebrae) having mobile connections (discs, ligaments).

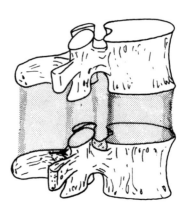

Movements of individual vertebrae are compounded such that the entire structure has considerable mobility in three dimensions. Type and extent of mobility varies with different spinal regions, depending on size and shape of vertebrae, and other factors.

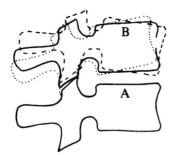

Let's look at what happens between two vertebrae during movement, assuming that the top vertebra (B) is mobile, while the bottom vertebra (A) is fixed.

Flexion ··

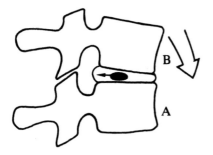

In **flexion**, B tilts toward the front. The disc is compressed anteriorly and expanded posteriorly, and its nucleus moves slightly toward the back.

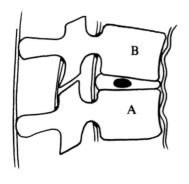

The superior articular facets slide on the inferior ones. The various parts of the vertebral arches are pulled apart, and the ligaments connecting these parts are stretched.

Extension

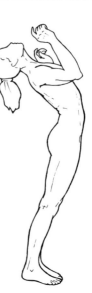

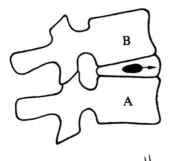

The opposite occurs in **extension**. B tilts toward the back. The disc is compressed posteriorly and expanded anteriorly, and its nucleus moves forward.

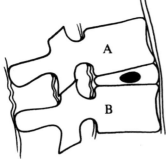

The articular facets are pressed together, the arches move closer together, and the posterior ligaments are relaxed. The anterior longitudinal ligament is stretched.

What happens in **lateral flexion**? Let's consider right side-bending as an example. The right sides of A and B move closer together, while the left sides move farther apart.

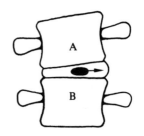

right   *left*

The disc is expanded (and its nucleus moves) to the left side. On the right side, the transverse processes and articular facets come closer together, and the associated ligaments are relaxed; the opposite occurs on the left side.

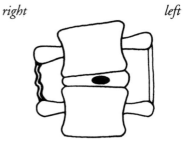

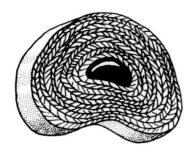

Within the annulus fibrosus of the disc, the fibers are oriented in alternating directions from one layer to the next.

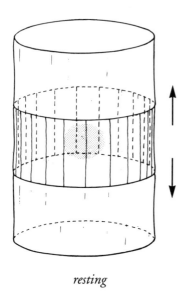

In **rotation**, therefore, some layers are stressed while others are relaxed. Because of the torsion (twisting) effect on the fibers, there is a reduction in overall height of the disc, and therefore a slight compression of the nucleus.

Externally, the transverse and spinous processes are moved apart by rotation, and their connecting ligaments are stretched.

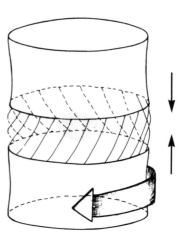

resting

rotation

The disc often receives pressure from the vertebral body above. The nucleus, because of its central location and gelatinous composition, tends to distribute the pressure it receives in every direction. Thus, the fibers of the annulus receive both horizontal and vertical pressures.

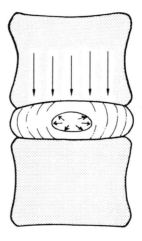

The disc performs its role as a shock absorber very efficiently as long as it is watertight. Unfortunately, due to the aging process and/or excessive "wear and tear," the disc may partially lose this property; i.e., cracks develop in the annulus through which the fluid of the nucleus can escape.

This condition is termed a herniated or **ruptured disc**. It happens most commonly as a result of chronic flexion movements, during which the nucleus moves toward the back and fluid can escape there. The fluid may then compress the nerve roots, e.g., the sciatic nerve which exits from the lumbar region, where pressures on the vertebral column are most intense. This situation, combined with chronic or sudden extreme tension on the posterior longitudinal ligament, can result in chronic lumbar backaches.

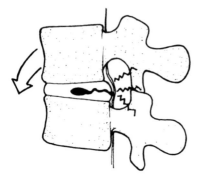

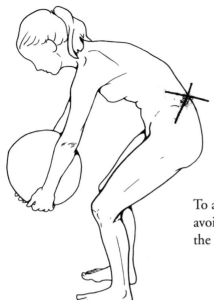

To avoid these problems, it is important to avoid "loaded" vertebral flexion, e.g., flexing the lumbar spine while lifting a heavy object.

Instead, keep the spine straight and flex at the hip and knee joints only.

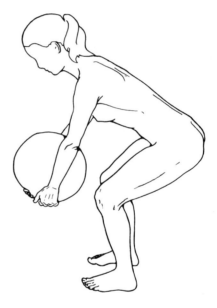

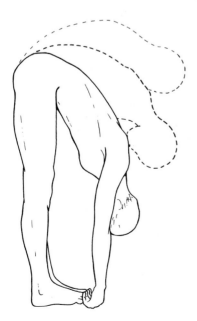

In fact, it is preferable to avoid loaded lumbar flexions in any type of physical exercise, even if you are not lifting any object.

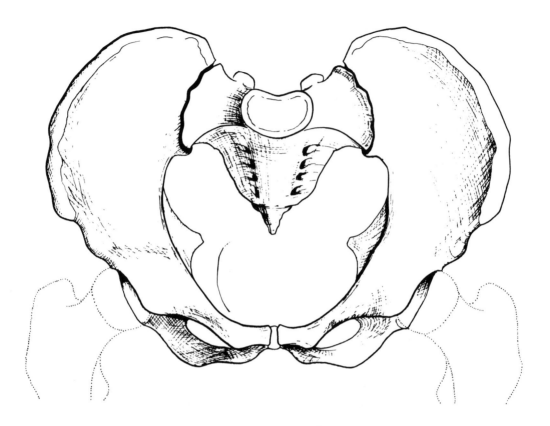

Pelvis

The pelvis (the word means "basin"), also called the pelvic ring, is a roughly cylindrical structure composed of several articulating or fused bones plus associated muscles (e.g., those making up the "pelvic floor") and ligaments (above).

It receives the weight of the upper body, and passes this weight on to the lower limbs via its articulations with the femurs. Conversely, it must absorb stresses from the lower limbs, e.g., in walking or jumping (right).

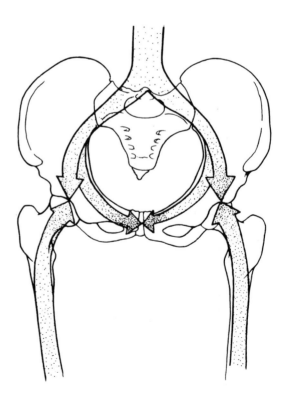

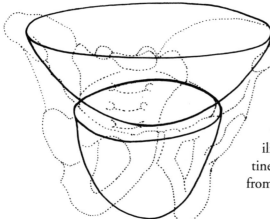

The pelvis is divided by an imaginary oblique plane into a superior **greater** ("false") **pelvis**, and an inferior **lesser** ("true") **pelvis**. If you look at a skeleton, you will see obvious "edges" on the sacrum and ilia (called the sacral promontory and iliopectineal lines respectively) that divide the greater from the lesser pelvis.

The superior and inferior openings of the lesser pelvis are called the pelvic inlet and pelvic outlet. During vaginal childbirth, the baby must pass through these narrow openings.

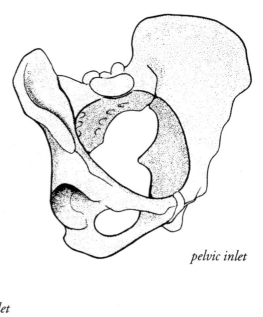

pelvic inlet

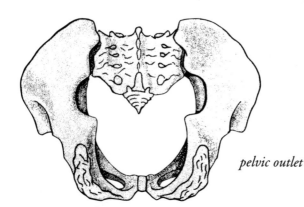

pelvic outlet

There are two **hip bones** or coxal bones. Each is composed of three elements, the **ilium**, **ischium**, and **pubis**, which are separated in the newborn but become fused by adulthood. At their junction, these three elements form a deep socket called the **acetabulum**, which articulates with the head of the femur. The ischium and pubis also form the circumference of a large opening called the **obturator foramen**.

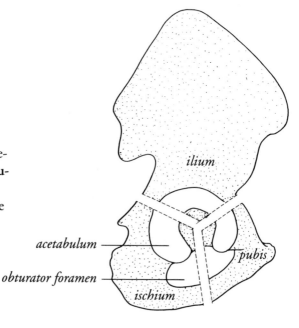

ilium

acetabulum

obturator foramen

pubis

ischium

The hip bone has two surfaces: lateral and medial.

Some important features
of the lateral surface are:

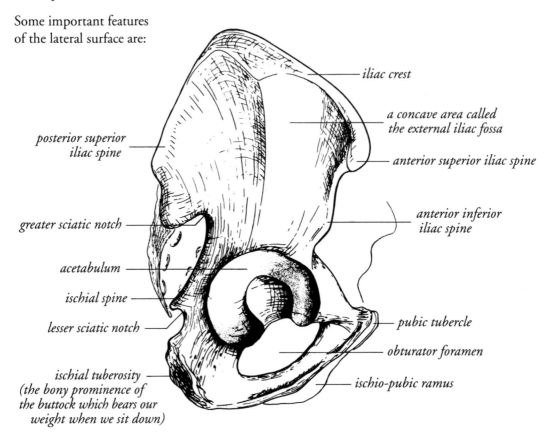

iliac crest

a concave area called
the external iliac fossa

posterior superior
iliac spine

anterior superior iliac spine

anterior inferior
iliac spine

greater sciatic notch

acetabulum

ischial spine

lesser sciatic notch

pubic tubercle

obturator foramen

ischio-pubic ramus

ischial tuberosity
(the bony prominence of
the buttock which bears our
weight when we sit down)

On the medial surface, we see:

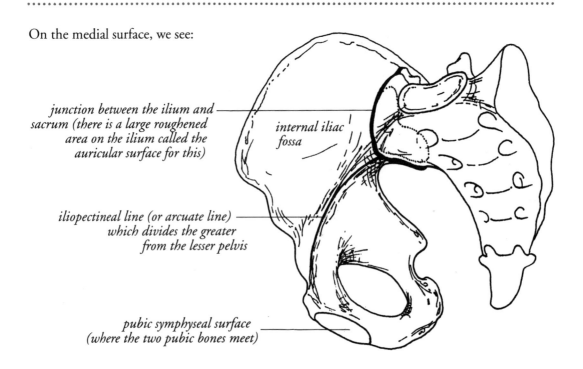

junction between the ilium and
sacrum (there is a large roughened
area on the ilium called the
auricular surface for this)

internal iliac
fossa

iliopectineal line (or arcuate line)
which divides the greater
from the lesser pelvis

pubic symphyseal surface
(where the two pubic bones meet)

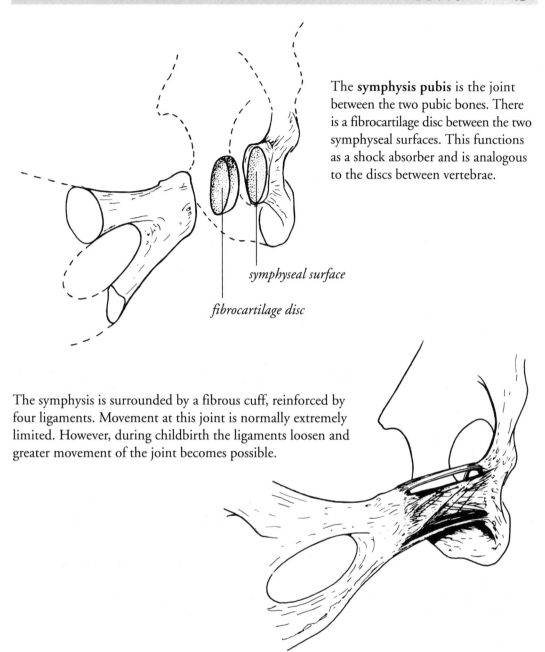

The **symphysis pubis** is the joint between the two pubic bones. There is a fibrocartilage disc between the two symphyseal surfaces. This functions as a shock absorber and is analogous to the discs between vertebrae.

symphyseal surface

fibrocartilage disc

The symphysis is surrounded by a fibrous cuff, reinforced by four ligaments. Movement at this joint is normally extremely limited. However, during childbirth the ligaments loosen and greater movement of the joint becomes possible.

The shape of the pelvis varies considerably in normal individuals. For example, the pelvic inlet may have a round shape (left), or be compressed in either direction (center, right).

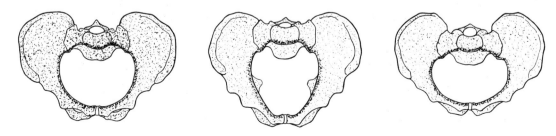

The curve of the sacrum may be more or less emphasized, and the hip bones more or less developed.

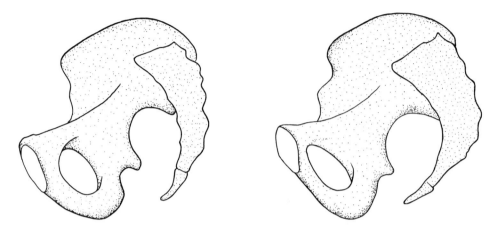

The gap between the two ischial tuberosities may be bigger or smaller.

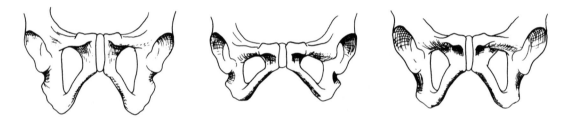

The sacral crest or posterior superior iliac spines may protrude in some individuals. This condition, combined with lack of "padding" by muscles and adipose tissue, can result in difficulty doing floor exercises, particularly rolling on a hard surface.

There are obvious differences in pelvic shape between males and females. Essentially, the male pelvis (left) is narrower and the female pelvis (right) is wider, with larger pelvic inlet and outlet, larger angle between the ischio-pubic rami, and more horizontal orientation of the ilia (for weight-bearing). These differences are all related to the ability of women to carry and deliver a child.

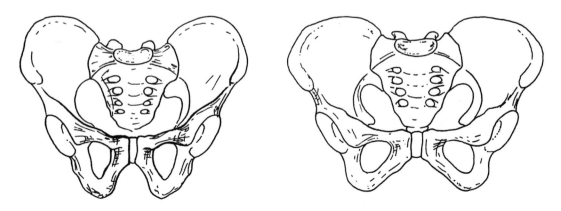

Sacrum

The sacrum is the posterior, wedge-shaped component of the pelvic ring, located between the two ilia. It is composed of five expanded, fused vertebrae (S1-S5) whose components are still easily visible.

The anterior surface is smooth and concave. There are four **transverse ridges**, which represent intervertebral discs. At the ends of each ridge are paired **anterior sacral foramina** (corresponding to intervertebral foramina), through which the anterior branches of the sacral nerves pass. The **sacral promontory** is the upper, anterior edge of S1. Together with the iliopectineal lines, it defines the boundary between the greater and lesser pelvis. The superior surface of S1, called the base of the sacrum, articulates with L5. The inferior surface of S5, which articulates with the coccyx, is called the apex of the sacrum.

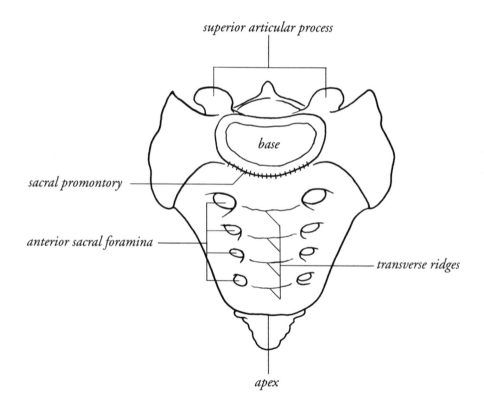

superior articular process

base

sacral promontory

anterior sacral foramina

transverse ridges

apex

The convex posterior surface (next page) is much rougher, with many bumps for attachment of muscles and ligaments. The **posterior sacral foramina** are continuous with the anterior foramina; posterior branches of the sacral nerves exit here. The **median sacral crest** and paired **lateral sacral crests** represent the spinous and transverse processes of the vertebrae. The superior **articular facets** meet the inferior facets of L5. The spinal cord enters through the **sacral canal**. There is also an inferior opening of the vertebral canal at S5 called the **sacral hiatus**, through which anesthetic agents may be injected. For this procedure, the **sacral cornua** (inferior articular processes of S5) provide useful landmarks.

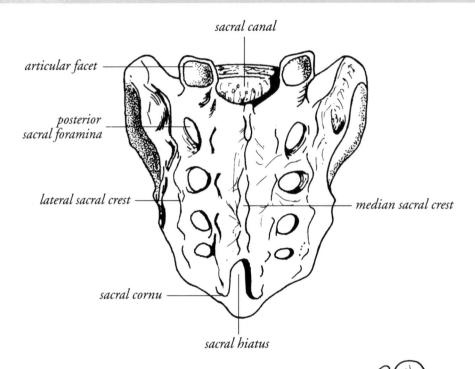

sacral canal

articular facet

posterior sacral foramina

lateral sacral crest

median sacral crest

sacral cornu

sacral hiatus

The **auricular surfaces** (this word means "ear-shaped") are best seen in side view. These articulate with the auricular surfaces of the ilia.

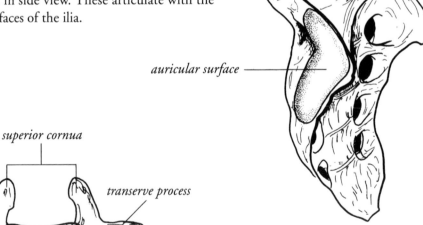

auricular surface

superior cornua

transerve process

The **coccyx**, the remnant of the tail from ancestral mammals, consists of three or four vestigial, fused vertebrae. The superior coccygeal cornua, corresponding to superior articular processes, are connected to the nearby sacral cornua by ligaments. The transverse processes are easily visible on the top coccygeal vertebra, but less so on the lower vertebra. The joint between the coccyx and sacrum is fibrous, and sometimes fused.

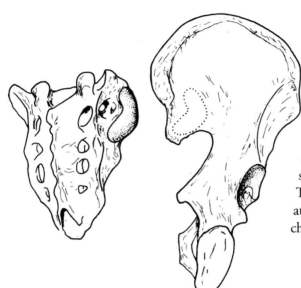

Sacroiliac joint

The auricular surfaces of the ilium and sacrum (particularly the lower parts) are slightly convex and concave, respectively. This arrangement allows slight movement at the sacroiliac joint, particularly during childbirth.

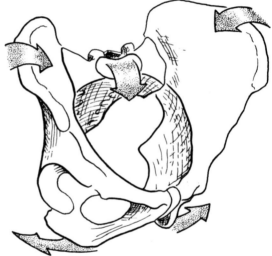

In one type of movement, which I will call "adduction" of the pelvis, the sacral base tilts anteroinferiorly, the sacral apex tilts postero-superiorly, the iliac "wings" are pulled medially, and the ischial tuberosities move laterally.

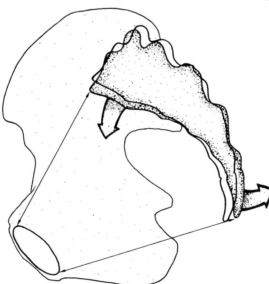

Thus, the pubic symphysis moves closer to the sacral base and farther from the sacral apex. The pelvic outlet increases in both dimensions, while the pelvic inlet decreases in both dimensions.

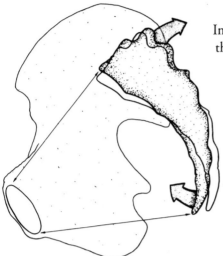

In the opposite movement ("abduction" of the pelvis), the pubic symphysis moves closer to the sacral apex and farther from the sacral base. The iliac wings move laterally, the ischia draw closer together, the pelvic outlet gets smaller, and the pelvic inlet gets larger.

"Abduction" of the pelvis occurs in the early stages of childbirth, when the baby begins to pass through the pelvic inlet. "Adduction" occurs during the final (expulsion) stage, when the baby exits through the pelvic outlet.

Each sacroiliac joints is reinforced by a capsule and a strong network of ligaments. The interosseus sacroiliac ligament is a thick, strong ligament connecting the rough area behind the auricular surface of the ilium to the superolateral sacrum.

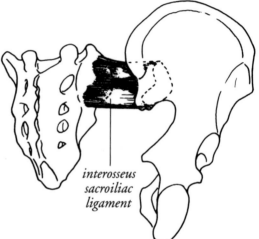

interosseus sacroiliac ligament

anterior sacroiliac ligament

The anterior sacroiliac ligament binds the sacrum to the medial surface of the ilium. Inferiorly, the sacrospinous and sacrotuberous ligaments connect the sacrum to the ischial spine and ischial tuberosity respectively. These ligaments tend to oppose "adduction" of the pelvis.

sacrospinous ligament

sacrotuberous ligament

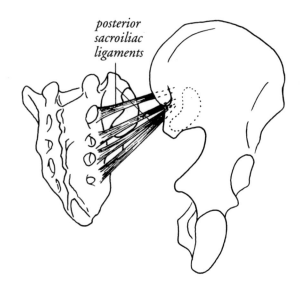

posterior sacroiliac ligaments

There are also a series of posterior sacroiliac ligaments connecting the ilium to the lateral sacral crest. These tend to oppose "abduction" of the pelvis.

Lumbar spine

The various regions of the vertebral column were introduced earlier. We now return to consider them in more detail. For brevity, we will use the word "spine" as a synonym for "vertebral column." Note that the meaning here is different than in terms such as "ischial spine" and "iliac spine," where we are simply referring to a small, sharp bony process.

The lumbar spine articulates with the sacrum, and is concave posteriorly (lordosis). In this region, the discs are about ⅓ the thickness of the vertebral bodies; this tends to increase mobility.

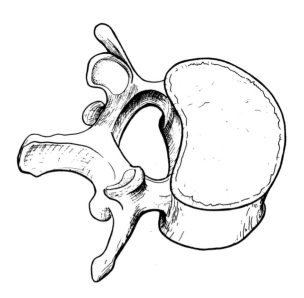

The bodies are large, and shaped like a lima bean. The transverse processes are long and have distinct tubercles on the end for muscle attachment.

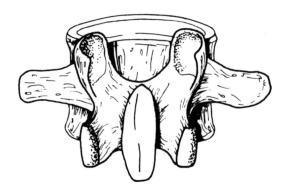

In contrast to other vertebral regions, the superior articular processes here are directed medially (rather than superiorly), and the inferior articular processes face laterally (rather than inferiorly).

Their articular facets are slightly concave and convex, respectively. The spinous processes are relatively short and massive, and project nearly straight posteriorly. These characteristics make rotation very difficult (below), but facilitate flexion, extension, and sidebending (right).

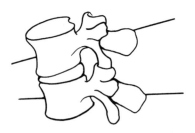

flexion

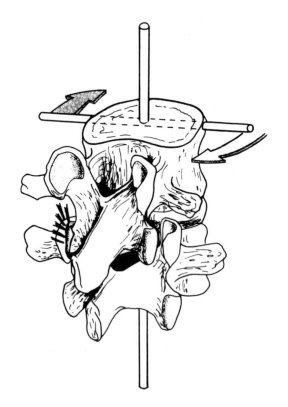

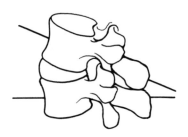

extension

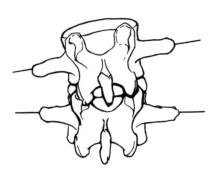

sidebending

Lumbosacral joint

The sacral base is tilted forward to a variable degree (considerably, in some individuals). Also, the body of L5 and the disc between L5 and S1 are slightly thicker anteriorly than posteriorly (this also applies to the L4/L5 joint). Thus, the lumbosacral joint is concave posteriorly.

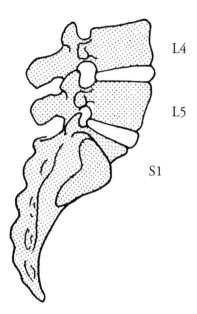

L4

L5

S1

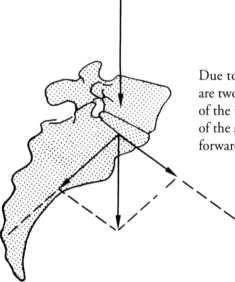

Due to the oblique orientation of the sacrum, there are two perpendicular forces resulting from the weight of the upper body at L5. One is directed along the axis of the sacrum, while the second tends to make L5 slide forward.

This second force can be significant if the sacral base is greatly tilted (right). In other words, L5 does not really "rest" on the sacral base, as many people think. It "wants" to slide forward. Notice that this tendency is opposed by the contact between the articular facets of S1 and the inferior articular processes of L5.

iliolumbar ligaments

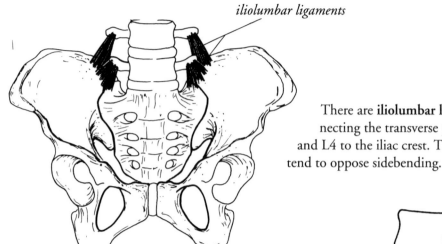

There are **iliolumbar ligaments** connecting the transverse processes of L5 and L4 to the iliac crest. These ligaments tend to oppose sidebending.

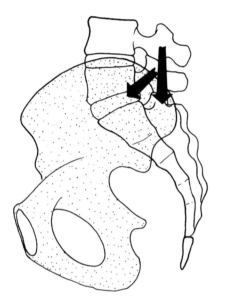

The ligament from L4 is oriented inferiorly, while that from L5 runs more anteroinferiorly. Thus, the L4 ligament is stretched and the L5 ligament relaxed during flexion (below left), while the reverse occurs during extension (below right).

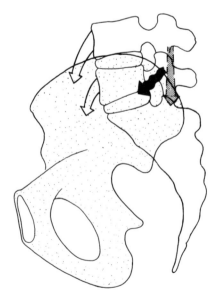

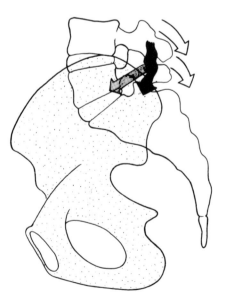

flexion

extension

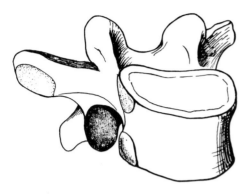

Thoracic spine

The bodies of the vertebrae in this region are roughly circular.

In contrast to the lumbar region, the thickness of the discs is only about ⅙ that of the bodies; this tends to limit range of motion (right). There are posterior facets or demifacets on the bodies for articulation with the heads of ribs. T1 has a whole facet near the top for articulation with the first rib, and an inferior demifacet which cooperates with a corresponding superior demifacet on T2 for articulation with the second rib. T2 through T8 each have superior and inferior demifacets. T9 has only a superior demifacet. T10 through T12 each have a whole facet.

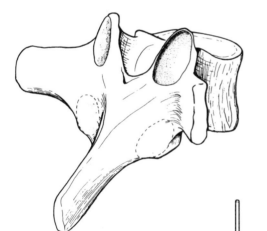

In this region, the facets of the articular processes are round and flat. They are oriented postero-superolaterally on the superior articular processes, and anteroinferomedially on the inferior processes. This arrangement permits flexion, extension, and sidebending.

These facets are located approximately on the circumference of a circle whose center would be the center of the vertebral body.

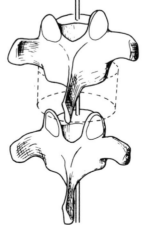

This facilitates rotation.

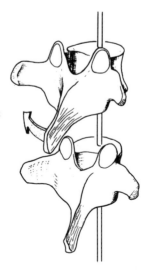

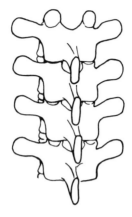

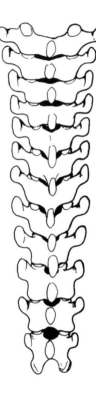

The laminae are flat, rectangular, and higher than wide. They are stacked on and overlap each other like tiles on a roof.

The spinous processes are elongated, laterally compressed, and (again in contrast to the lumbar region) directed inferiorly. These characteristics help prevent hyperextension.

The transverse processes decrease in length from top to bottom (right), and those of T1 through T10 have anterior facets for articulation with the tubercles of ribs.

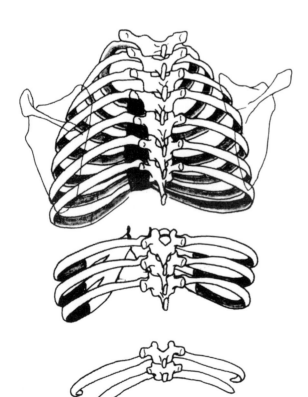

The attachment to the rib cage tends to limit mobility of the thoracic spine. This is particularly true for T1 through T7, whose corresponding ribs (the "true ribs") are connected directly to the sternum by short pieces of cartilage which allow little mobility.

Ribs 8, 9, and 10 (the "false" ribs) have longer costal cartilages which attach to the cartilage of rib 7 rather than directly to the sternum. The mobility of T8, T9, and T10 is correspondingly greater.

Ribs 11 and 12 (the "floating" ribs) have no anterior attachment at all, so the mobility of T11 and T12 is greatest.

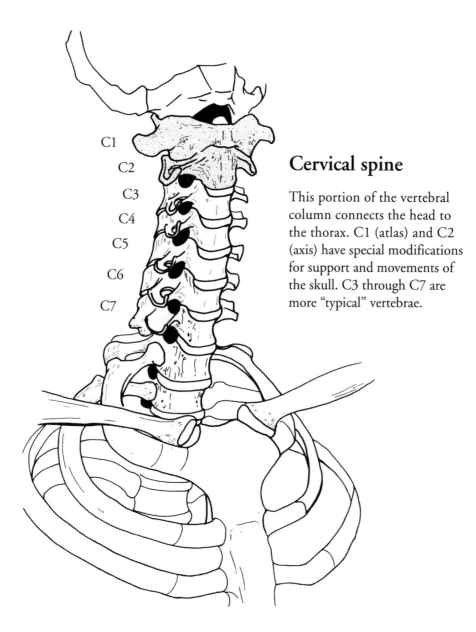

C1
C2
C3
C4
C5
C6
C7

Cervical spine

This portion of the vertebral column connects the head to the thorax. C1 (atlas) and C2 (axis) have special modifications for support and movements of the skull. C3 through C7 are more "typical" vertebrae.

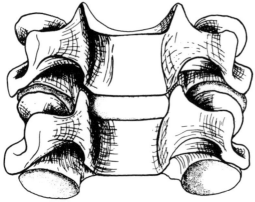

The bodies of cervical vertebrae are small (relative to the thoracic and lumbar regions), and the discs about ⅓ as thick as the bodies. Both these factors tend to increase mobility. Sidebending is somewhat restricted by the rectangular shape of the bodies as seen in cross-section.

In C3-C7, the superior surface of the body is concave transversely, with two projecting "lips" on either side. The anterior aspect of this surface is more convex, and tilted slightly forward. The inferior surface of the body of the adjacent vertebra is correspondingly shaped. This structure gives stability to the joint while still allowing good mobility.

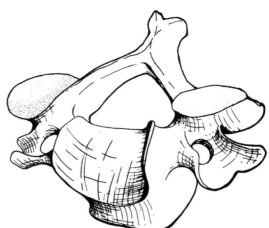

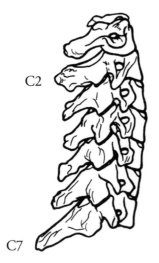

C2

C7

The spinous processes of C2 and C7 are relatively long. That of C7 can be easily palpated and seen externally, and provides a good landmark.

The short spinous processes of the other cervical vertebrae allow good extension (right). The spinous processes of C2-C6 are bifid, i.e., they have a small notch at the tip. A strong fibrous band called the **nuchal ligament**, which helps support the weight of the head, runs along these notches and inserts on the external occipital crest.

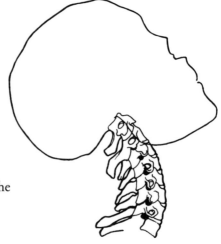

The cervical transverse processes are shaped differently from those of the thoracic spine. They are short and broad, and tend to limit sidebending when they come in contact.

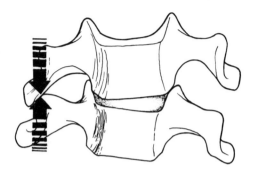

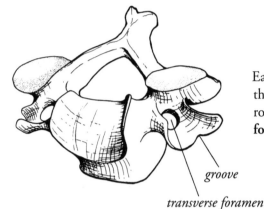

groove

transverse foramen

Each has two roots: one from the side of the body, the other from the pedicle. As they merge, the two roots form a medial opening called the **transverse foramen**, and a small lateral groove.

The vertebral artery and vein pass through the transverse foramina of C1-C6. Spinal nerves pass through the lateral grooves. Thus, correct alignment of the cervical spine is important for protection of these soft tissues.

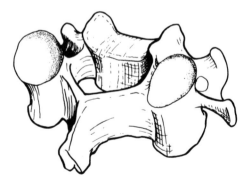

The superior articular facets face posterosuperiorly, and the inferior facets anteroinferiorly, at an angle of 45°.

Sidebending of the cervical spine is always accompanied by a certain amount of rotation. For example, during left sidebending, the left facet move inferiorly and slightly posteriorly, while the right facet moves superiorly and slightly anteriorly. The combination of these two movements results in slight left rotation.

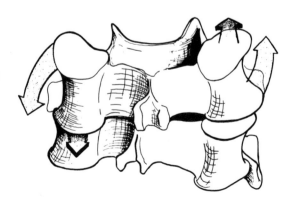

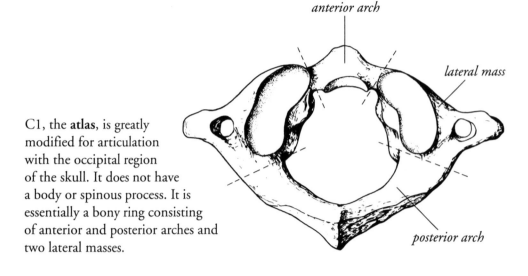

anterior arch

lateral mass

posterior arch

C1, the **atlas**, is greatly
modified for articulation
with the occipital region
of the skull. It does not have
a body or spinous process. It is
essentially a bony ring consisting
of anterior and posterior arches and
two lateral masses.

The transverse processes, containing the transverse foramina, project from the sides of the lateral masses. A transverse ligament connects the two lateral masses and divides the large central cavity into a posterior portion (the vertebral foramen, for passage of the spinal cord) and smaller anterior portion (which accommodates the dens of the axis).

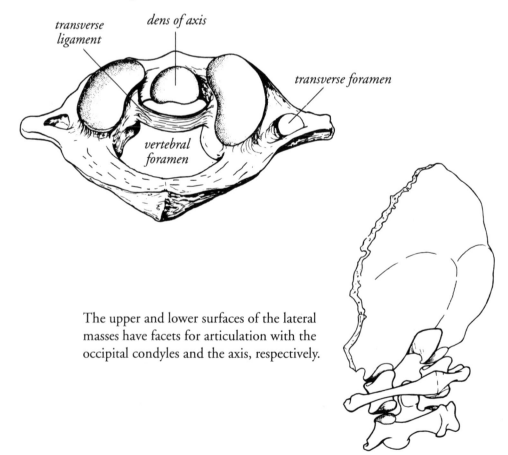

transverse
ligament

dens of axis

transverse foramen

vertebral
foramen

The upper and lower surfaces of the lateral
masses have facets for articulation with the
occipital condyles and the axis, respectively.

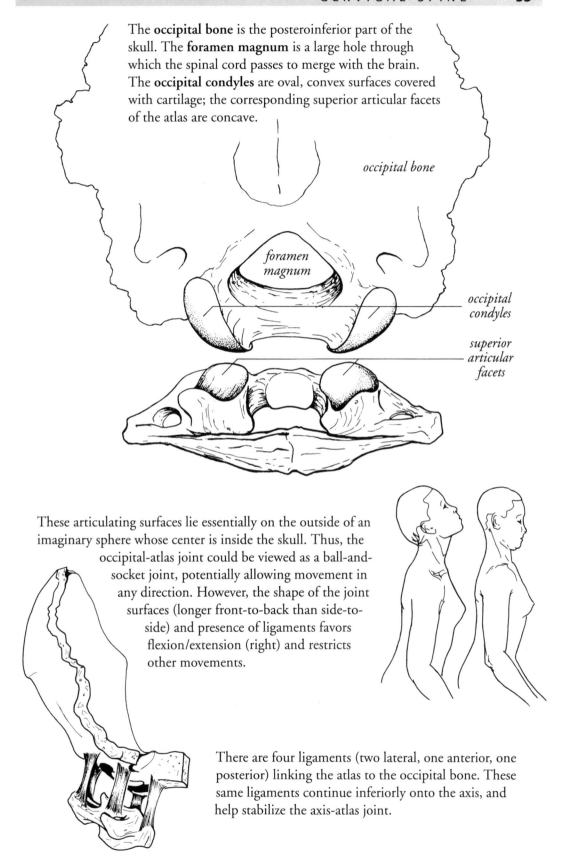

The **occipital bone** is the posteroinferior part of the skull. The **foramen magnum** is a large hole through which the spinal cord passes to merge with the brain. The **occipital condyles** are oval, convex surfaces covered with cartilage; the corresponding superior articular facets of the atlas are concave.

occipital bone

foramen magnum

occipital condyles

superior articular facets

These articulating surfaces lie essentially on the outside of an imaginary sphere whose center is inside the skull. Thus, the occipital-atlas joint could be viewed as a ball-and-socket joint, potentially allowing movement in any direction. However, the shape of the joint surfaces (longer front-to-back than side-to-side) and presence of ligaments favors flexion/extension (right) and restricts other movements.

There are four ligaments (two lateral, one anterior, one posterior) linking the atlas to the occipital bone. These same ligaments continue inferiorly onto the axis, and help stabilize the axis-atlas joint.

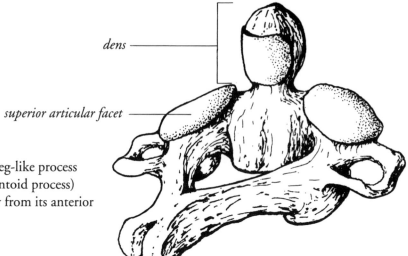

dens

superior articular facet

C2, the **axis**, has a peg-like process called the **dens** (odontoid process) projecting superiorly from its anterior side.

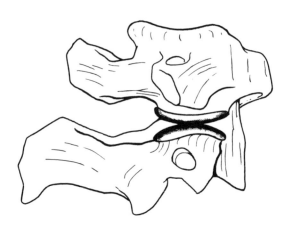

The dens actually represents the missing body of C1, which becomes attached to the body of C2 during embryogenesis. The superior articular facets of C2 are convex. Because the corresponding inferior facets of C1 are also convex, and because the disc between these two vertebrae is poorly-developed, this is not a tight-fitting joint.

Considerable movement, especially rotation, is possible. The dens fits anteriorly against the anterior arch of C1, and posteriorly against an "articular surface" of the transverse ligament of C1. Thus, the pivot joint of C1-C2 consists of a ring-like structure rotating around the dens.

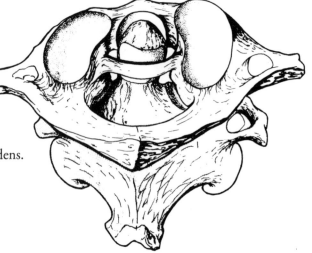

Rotation of C1 on C2 is accompanied by some lateral gliding of C2 which helps preserve the integrity of the spinal canal. These movements involve four joints: two at the lateral articular facets, and two at the front and back of the dens.

The axis of rotation can pass either through the dens...

...or one of the articular facets.

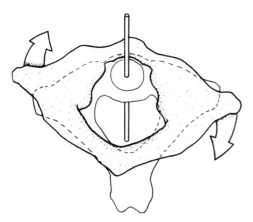

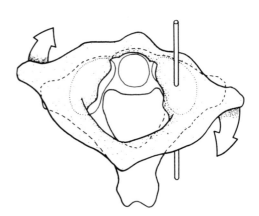

There are anterior (not shown here) and posterior ligaments between C2 and C1, as well as ligaments (shown here as arrows) connecting the occiput to the body and dens of C2.

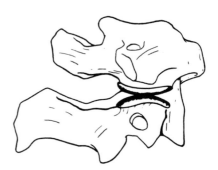

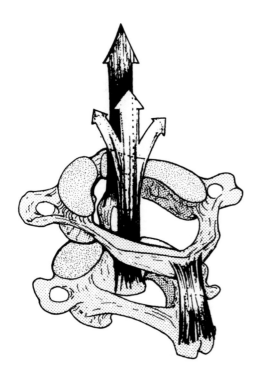

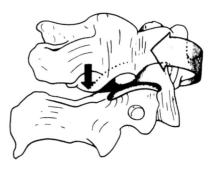

Due to the convexity of the articular facets of both C1 and C2, C1 tends to move slightly closer to C2 during rotation.

Posterior muscles of trunk

We turn our attention now from the bones to the muscles of the trunk, beginning with the posterior ones.

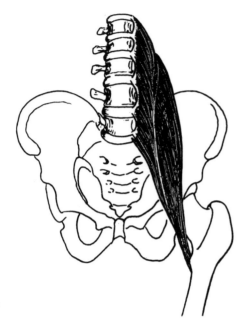

Lumbar muscles

Psoas major originates from the bodies and transverse processes of L1-L5, passes anterior to the pelvis, and inserts on the lesser trochanter of the femur. Depending on which end is fixed, it can act on either the thigh or vertebral column.

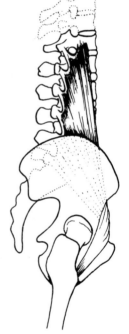

Here, we shall consider what happens when the thigh (i.e., hip joint) is fixed. Because of the orientation of its fibers (obliquely anteroinferior), bilateral contraction of the psoas is generally considered to result in flexion of the lumbar spine, or increasing its curvature (lordosis). However, electromyographic recordings taken from moving subjects suggest a paradoxical action.

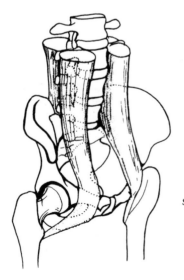

The psoas, in combination with the posterior transverso-spinalis muscles, forms a system of four muscular bundles arranged around the lumbar spine. By contracting together, these four bundles can act to erect (straighten) the lumbar spine, rather than increasing lordosis.

Unilateral contraction of the psoas results in ipsilateral (i.e., on the same side as the contracting muscle) sidebending or contralateral (i.e., to the side opposite the contracting muscle) rotation of the lumbar spine.

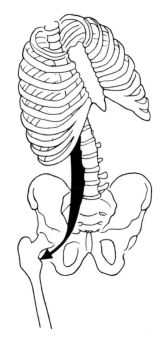

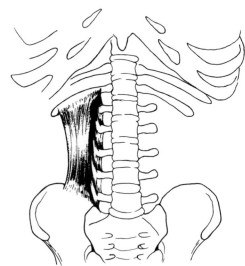

Quadratus lumborum originates from the posterior iliac crest, and inserts on rib 12 and the transverse processes of L1-L5. It is composed of both vertical and oblique fibers.

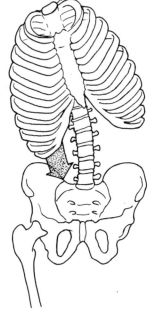

When the pelvis is fixed, contraction of this muscle causes sidebending of the lumbar spine and ribcage.

When the ribs and spine are fixed, it raises the pelvis on one side.

Deep back muscles

Trunk muscles tend to be arranged in several layers, with the deepest ones consisting of small bundles passing from one vertebra to another.

The **intertransverse** muscles connect one transverse process to the next, posterior to the intertransverse ligament. *Action:* sidebending.

The **interspinalis** muscles connect adjacent spinous processes, on either side of the ligament. *Action:* extension.

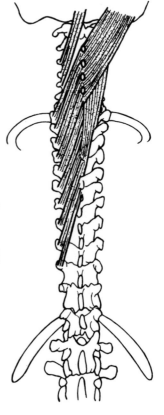

The **transversospinalis** muscles run superomedially from transverse process to spinous process. There are three subdivisions. The **semi-spinalis** span five or six vertebral levels. The **multifidus** span three levels and are deep to semispinalis. The **rotatores** span a single level and are deep to the other two.

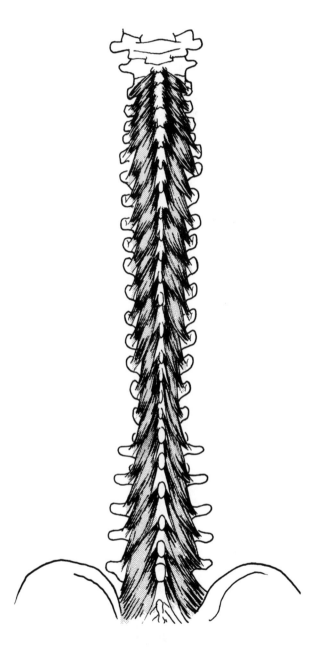

When viewed from behind, the transverso-spinalis form a "chevron"-like pattern.

Depending on whether they are contracting on both sides or one side only, and on what other muscles are doing, they can assist in extension...

...sidebending, or rotation.

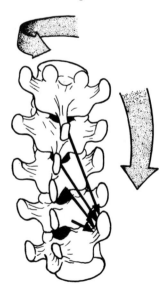

Electromyographic studies have shown that the activity and importance of the transversospinalis are variable depending upon spinal level, particularly in terms of "elongating" (straightening) the spine.

It is important around T5/T6, where the posterior convexity of the thoracic spine is most pronounced.

It is less important around C5 and L3, where the posterior concavity of the cervical and lumbar regions is most pronounced, and the anterior neck muscles and psoas (respectively) play a greater role.

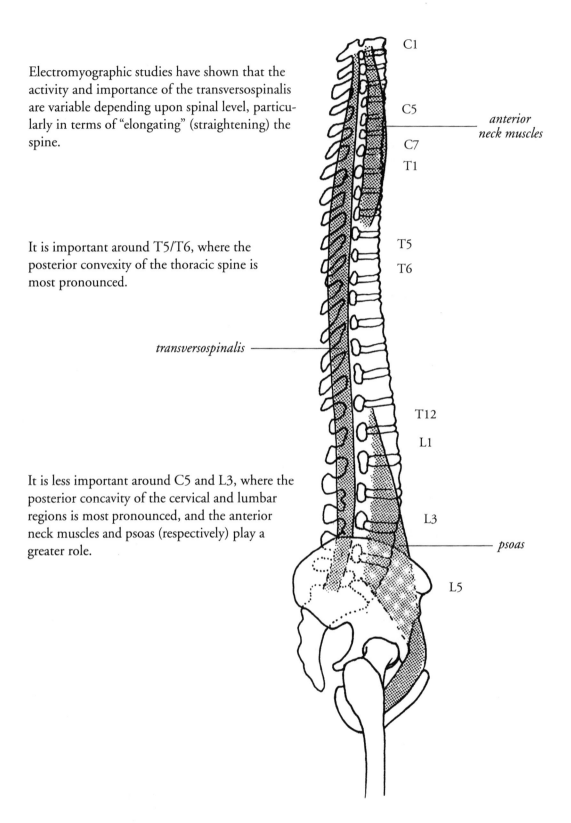

C1

C5

C7

T1

T5

T6

T12

L1

L3

L5

anterior neck muscles

transversospinalis

psoas

Deep neck muscles

These are analogous to the transversospinalis but insert on the occiput. **Rectus capitis posterior minor** runs from the posterior arch of C1 to the inferior occipital ridge. **Rectus capitis posterior major** originates from the spinous process of C2 and inserts just lateral to the minor.

rectus capitis posterior

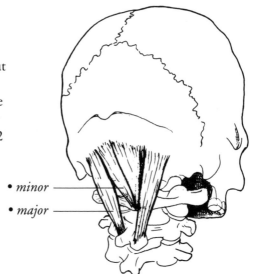

- *minor*
- *major*

Obliquus capitis superior originates from the transverse process of C1 and inserts on the occiput lateral to r.c.p. major, just posterior to the mastoid process of the temporal bone. These three muscles help produce extension at the C1/C2 joint if they contract bilaterally. **Obliquus capitis inferior** runs from the spinous process of C2 to the transverse process of C1.

obliquus capitis

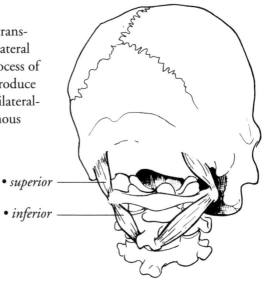

- *superior*
- *inferior*

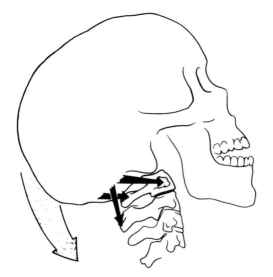

Action: extension, sidebending, or slight rotation of C1 on C2

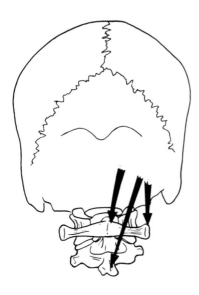

The three muscles that insert on the occiput, by contracting unilaterally (on one side only), bend the head to that side. Obliquus capitis superior, since it has the most lateral insertion, does this most effectively.

The two rectus capitis muscles, by contracting on the right, can produce right rotation.

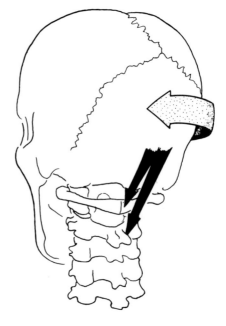

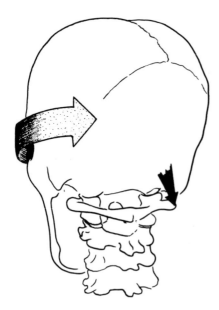

Obliquus capitis superior, because of its orientation, can produce left rotation by contracting on the right.

These deep muscles have limited lever action because of their small size, but allow great precision of movement. In cooperation with the anterior neck muscles, they regulate the correct orientation of the head on the neck.

Intermediate back and neck muscles

The group of posterior muscles shown on this page forms a layer superficial to those described above. It is sometimes referred to collectively as the sacrospinalis or erector spinae.

There are three components: iliocostalis (most lateral), longissimus, and spinalis (most medial). Each of these is further divided into three subcomponents.

Longissimus capitis originates from the transverse processes of the upper thoracics and lower cervicals, and inserts on the mastoid process of the temporal bone.

iliocostalis cervicis

longissimus capitis

longissimus cervicis

Longissimus thoracis is visible on the right side of the illustration. It originates from the lumbar transverse processes and inserts on the thoracic transverse processes and posterior aspect of ribs 9 and 10. It fills the groove formed where the thoracic vertebrae meet the ribs.

iliocostalis thoracis

longissimus thoracis

The iliocostalis has been "cut away" on the right side of the illustration to reveal the longissimus, but is left intact on the left side. **Iliocostalis lumborum** originates from the iliac crest via the lumbar fascia (a sheet-like structure made of dense connective tissue), and inserts on the lower ribs. **Iliocostalis thoracis** runs from the lower six to the upper six ribs. **Iliocostalis cervicis** runs from the upper six ribs to the transverse processes of the lower cervicals.

iliocostalis lumborum

lumbar fascia

Longissimus cervicis runs from the transverse processes of the upper thoracic vertebrae to those of C2-C6.

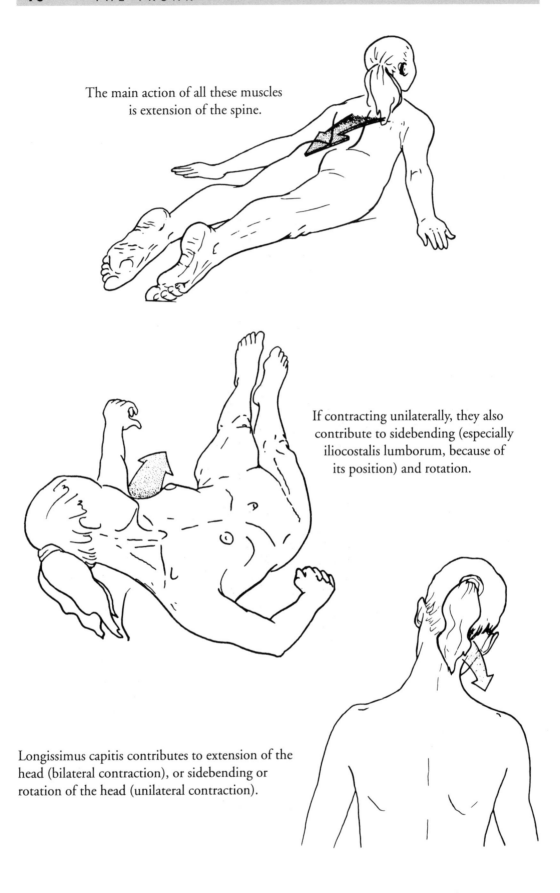

The main action of all these muscles
is extension of the spine.

If contracting unilaterally, they also
contribute to sidebending (especially
iliocostalis lumborum, because of
its position) and rotation.

Longissimus capitis contributes to extension of the
head (bilateral contraction), or sidebending or
rotation of the head (unilateral contraction).

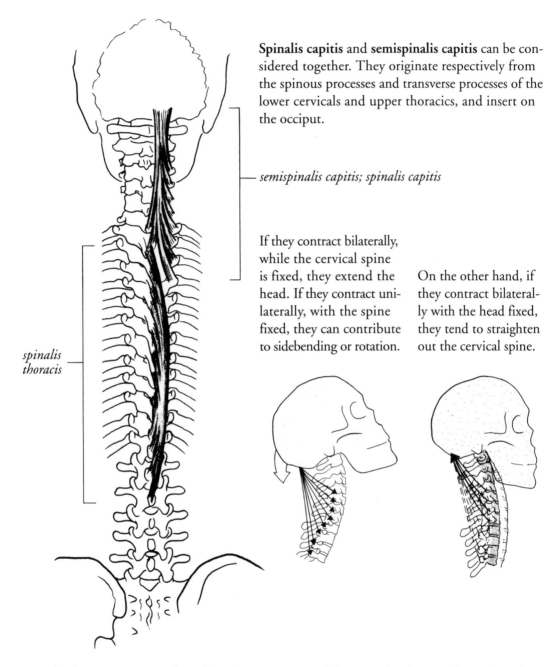

spinalis thoracis

semispinalis capitis; spinalis capitis

Spinalis capitis and **semispinalis capitis** can be considered together. They originate respectively from the spinous processes and transverse processes of the lower cervicals and upper thoracics, and insert on the occiput.

If they contract bilaterally, while the cervical spine is fixed, they extend the head. If they contract unilaterally, with the spine fixed, they can contribute to sidebending or rotation.

On the other hand, if they contract bilaterally with the head fixed, they tend to straighten out the cervical spine.

Spinalis thoracis originates from the spinous processes of the upper lumbars and lower thoracics, and inserts on those of the upper thoracics. Because of its medial location, it extends the spine but plays no part in sidebending or rotation.

All the deep and intermediate back and neck muscles described so far function constantly (and usually subconsciously) to maintain the correct position of the head and spine while we are walking, sitting, etc. They are physiologically adapted to work in relays, i.e., take turns contracting, so that they do not fatigue under normal conditions. We typically become aware of them only when something goes wrong.

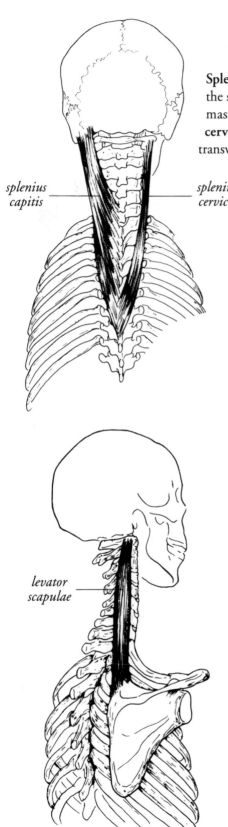

*splenius
capitis*

*splenius
cervicis*

*levator
scapulae*

Splenius capitis originates from the nuchal ligament and the spinous processes of C7 through T6. It inserts on the mastoid process and adjacent occipital bone. **Splenius cervicis** runs from the spinous process of T5-T7 to the transverse processes of C1-C3.

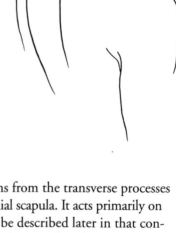

Contracting bilaterally, these muscles extend the head and cervical spine. Contracting unilaterally, they cause sidebending and rotation toward the contracting side.

Levator scapulae runs from the transverse processes of C1-C4 to the medial scapula. It acts primarily on the scapula, and will be described later in that context. However, when the scapula is fixed, its actions can reinforce those of the splenius cervicis.

Serratus posterior are flat muscles which assist in respiration. The superior portion runs inferolaterally from the spinous processes of C7 and T1-T3 to the posterior surfaces of ribs 2-5 near their angles. It elevates the ribs and thereby aids inspiration.

serratus posterior superior

The inferior portion runs superolaterally from the spinous processes of T11-T12 and L1-L2 to ribs 9-12. It depresses these ribs and aids expiration.

serratus posterior inferior

The following three muscles act primarily on the shoulder joint and will be discussed later in that context. However, when the shoulder is fixed, they can also act on the spine. The **rhomboids** (major and minor) run from the spinous processes of C7 and T1-5 to the medial scapula. When the scapula is fixed, their contraction pulls these vertebrae laterally. This effect is sometimes utilized by physiotherapists for vertebral re-alignment (e.g., in scoliotic cases).

rhomboids (major & minor)

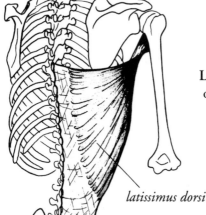

Latissimus dorsi has a very broad origin extending from T7 to the sacrum. It inserts on the humerus.

latissimus dorsi

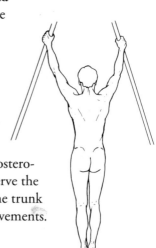

By wrapping around the postero-inferior trunk, it helps preserve the structural integrity of the trunk during certain movements.

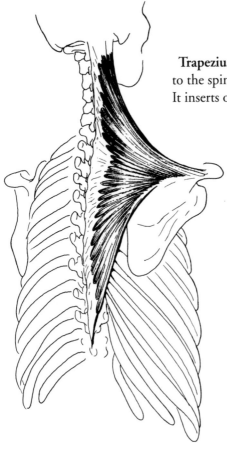

Trapezius also has a broad origin, extending from the occiput to the spinous processes of all cervical and thoracic vertebrae. It inserts on the superior scapula and clavicle.

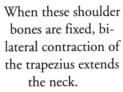

When these shoulder bones are fixed, bilateral contraction of the trapezius extends the neck.

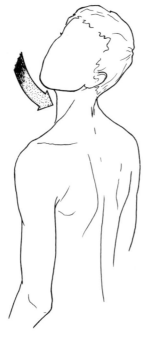

Unilateral contraction of its superior portion can assist in ipsilateral sidebending, or contralateral rotation of the head, due to the orientation of the fibers.

Anterior neck muscles

Longus colli is a deep muscle consisting of three portions. The longitudinal portion runs from the bodies of C2 through T3 to the bodies of C4-C7. The oblique superior portion runs from the anterior arch of C1 to the transverse processes of C3-C6. The oblique inferior portion runs from the bodies of T1-T3 to the transverse processes of C5-C7.

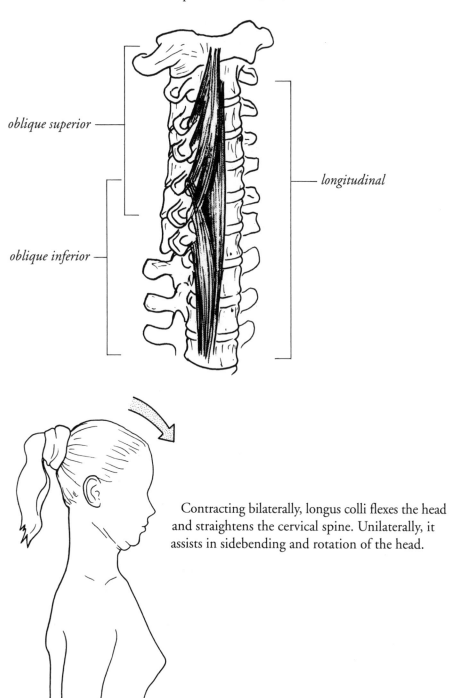

oblique superior

oblique inferior

longitudinal

Contracting bilaterally, longus colli flexes the head and straightens the cervical spine. Unilaterally, it assists in sidebending and rotation of the head.

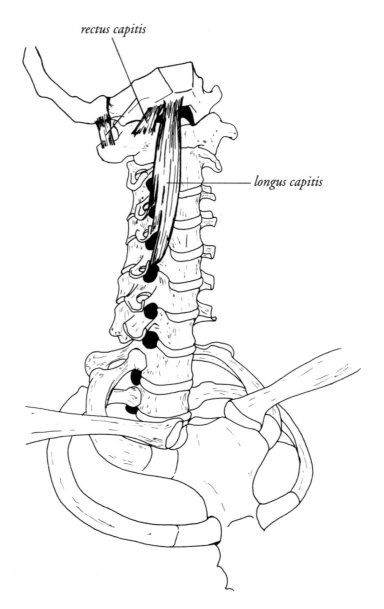

rectus capitis

longus capitis

Rectus capitis is a small muscle running from C1 to the occiput. It has an anterior and a lateral portion. Depending on whether contraction is bilateral or unilateral, it helps in flexion, sidebending, or rotation of the head. **Longus capitis** originates from the transverse processes of C3-C6 and inserts on the occiput just anterior to the rectus. It helps to straighten the upper cervical spine and flex the head (bilateral action), or to sidebend the head (unilateral action).

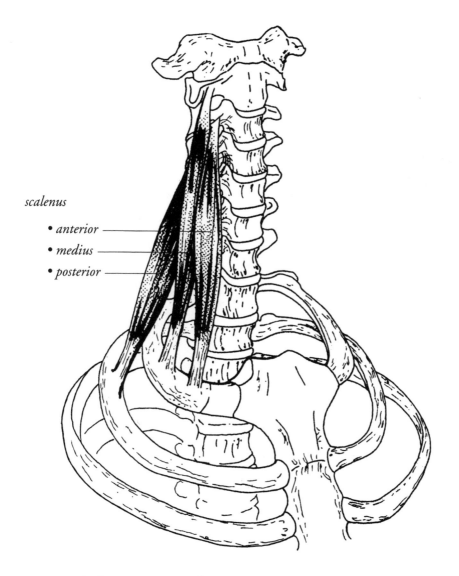

scalenus

- *anterior* ————
- *medius* ————
- *posterior* ————

Scalenus anterior and **scalenus medius** originate from the transverse processes of C3-C6 and C2-C7 respectively, and insert close together on the anterior part of rib 1. **Scalenus posterior** runs from the transverse processes of C4-C6 to the lateral surface of rib 2.

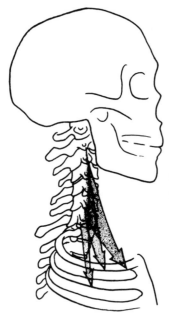

 The orientation of scalenus posterior is almost vertical, whereas scalenus anterior runs obliquely forward.

When the ribs are fixed, unilateral contraction of the scalenes (especially posterior) produces sidebending of the cervical spine; medius and anterior also produce some contralateral rotation.

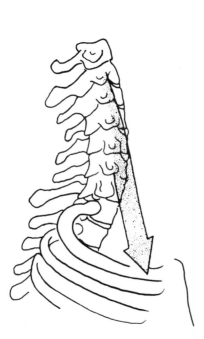

Bilateral contraction of medius and anterior accentuates the curvature of the spine. When the cervical (and upper thoracic) spine is fixed, contraction of the scalenes elevates ribs 1 and 2, assisting in inspiration.

Longus colli and longus capitis play an important role in stabilizing the cervical spine during this action.

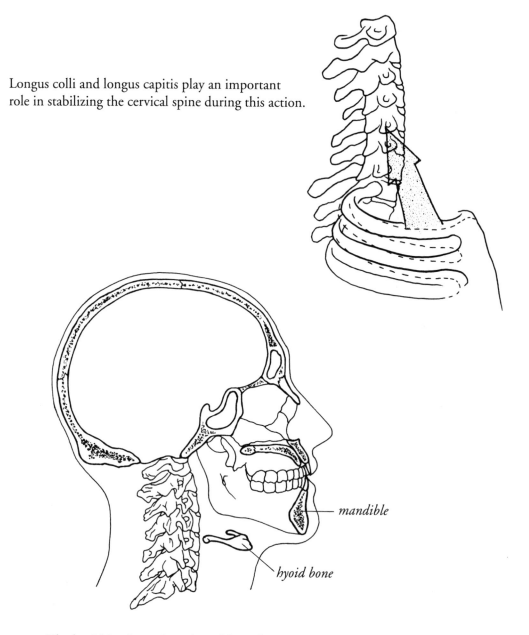

mandible

hyoid bone

The **hyoid** is a horseshoe-shaped bone located between the mandible and larynx, beneath the base of the tongue. It is unique in that it does not articulate with any other bone. Rather, it is held in place by numerous muscles (plus a ligament which suspends it from the styloid process of the temporal bone). It functions mainly as an origin for muscles of the tongue, which assist in the complex actions of speech, swallowing, etc. The infrahyoid muscles (sternohyoid, thyrohyoid, omohyoid) insert on the hyoid and have the origins below it (scapula, clavicle, sternum). The suprahyoid muscles (hyoglossus, geniohyoid, mylohyoid, digastric, stylohyoid) originate above the hyoid. These muscles are primarily responsible for stabilizing the hyoid, but some of them play a small role in flexing the head.

Sternocleidomastoid (SCM) is the largest and most important anterior neck muscle. It can be clearly seen externally when the head is turned (right).

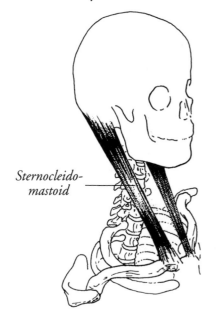

Sternocleido-mastoid

As revealed by its name, it has dual origins on the sternum and clavicle (near their junction), and inserts on the mastoid process.

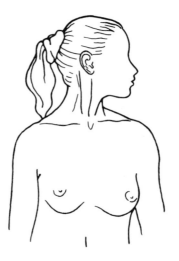

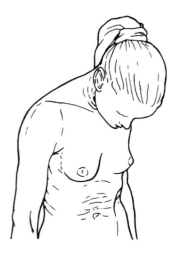

When the skull is fixed, SCM elevates the sternum and clavicle, and thereby assists in inspiration.

When the thoracic cage is fixed, unilateral contraction of SCM causes ipsilateral sidebending and contralateral rotation of the head.

Bilateral contraction results mainly in flexing of the head and cervical spine (you can easily verify this by placing your hand on SCM and trying to flex your head against some resistance). However, due to the oblique orientation of this muscle, and its insertion on the mastoid process, it also plays a role in extension.

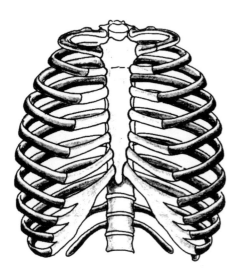

Thoracic cage

The thoracic cage consists of the thoracic vertebrae (posteriorly), sternum (anteriorly), and ribs (laterally).

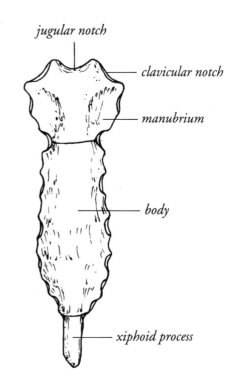

The **sternum,** or breastbone, is fairly flat. It consists of three parts: the manubrium, body, and xiphoid process. The **manubrium** and body are embryologically distinct but become fused by adulthood. The xiphoid process consists of hyaline cartilage in children and does not ossify completely until about age 40. The manubrium has a prominent jugular notch (easily seen and palpated externally) at the superior end, and two notches on either side of this for articulation with the clavicles. Farther down are the notches for the first pair of ribs. The notch for rib 2 is at the junction between the manubrium and body. Ribs 3 through 7 all articulate with the body.

The **ribs** are elongated, flattened, and twisted bones. At the posterior end of each rib is a head with two facets for articulation with the bodies of thoracic vertebrae. The neck is a constricted portion next to the head. Next is the tubercle, which includes an articular part (connected to the transverse process of a thoracic vertebra) and nonarticular part (for ligament attachment). The body, or shaft, is long and curved. The anterior end is joined to the costal cartilage.

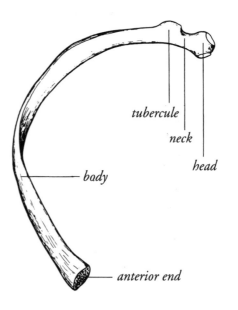

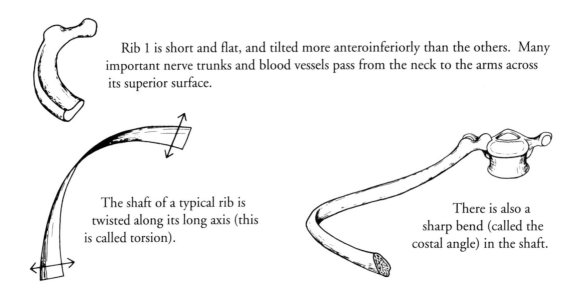

Rib 1 is short and flat, and tilted more anteroinferiorly than the others. Many important nerve trunks and blood vessels pass from the neck to the arms across its superior surface.

The shaft of a typical rib is twisted along its long axis (this is called torsion).

There is also a sharp bend (called the costal angle) in the shaft.

As a result, the lower border of a rib cannot be placed on a flat surface. The axial torsion and angle are significant in the movements of respiration. The bent ribs are under tension, and can be seen to move apart when their connection to the sternum is cut during a surgical operation.

Most ribs articulate with two thoracic vertebrae at three points, as mentioned above: the two facets on the head contact the demifacets on the vertebral bodies, and the tubercle contacts the transverse process.

However, ribs 1 (the small, uppermost rib) and 11-12 (floating ribs) contact a whole facet on the corresponding vertebra rather than two demifacets, and do not contact transverse processes. These rib-vertebra joints are stabilized by several small ligaments (below).

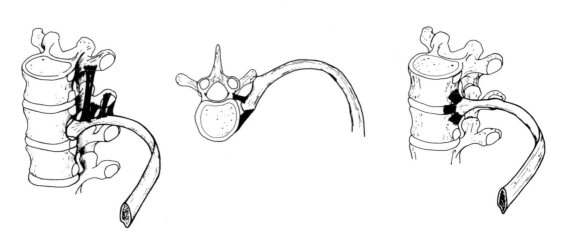

Anteriorly, the elasticity of the thoracic cage is increased by the presence of the **costal cartilages** (right). Mobility is greater for the false ribs (which attach only to the cartilage of rib 7) than for the true ribs (which attach directly to the sternum), and greatest of all for the floating ribs (which have no anterior attachment).

The movements of a rib can be compared to those of a bucket handle. As the rib moves up or down, the diameter of the thoracic cage is changed.

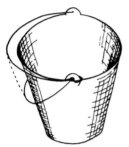

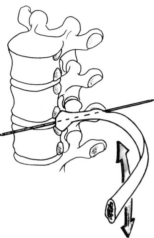

Posteriorly, the rib pivots on an axis passing through the two joints where its head and tubercle contact the body and transverse process of the vertebra.

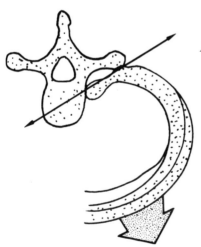

At different levels, the spatial relationship of these two joints changes because the shape of the vertebra changes. For the superior thoracic vertebrae, the axis is directed more laterally, and elevation of the rib increases the anterior diameter of the thoracic cage.

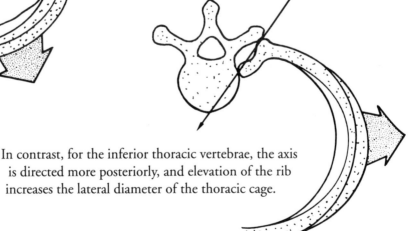

In contrast, for the inferior thoracic vertebrae, the axis is directed more posteriorly, and elevation of the rib increases the lateral diameter of the thoracic cage.

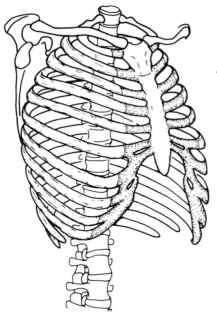

The anterior ends of the ribs easily accommodate these movements because of the great flexibility of the costal cartilages. However, this flexibility can decrease with age, thus reducing overall mobility.

During **inspiration** (inhalation), the ribs are elevated. Therefore, the diameter of the upper thoracic cage is increased in an anterior direction, while that of the lower cage is increased in a lateral direction.

The reverse occurs during **expiration** (exhalation), when the ribs move down. The costal cartilages undergo some torsion (twisting on their own axis) on inspiration and return to their normal shape on expiration.

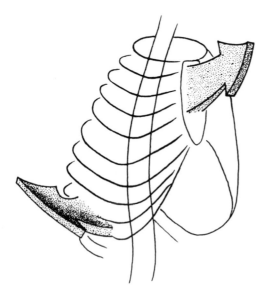

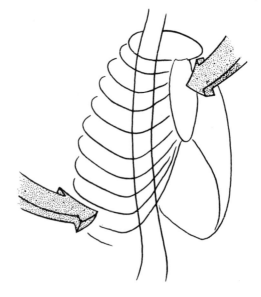

Obviously, spinal movements also affect the orientation of the ribs. The ribs in front move closer together during thoracic flexion...

...and farther apart during extension.

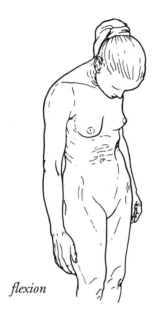

flexion

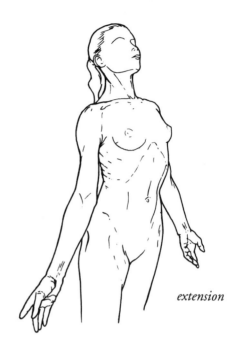

extension

In right sidebending, the ribs move closer together on the right side and farther apart on the left side.

In right rotation, the ribs move posteriorly on the right and anteriorly on the left.

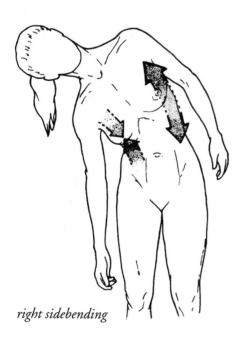

right sidebending

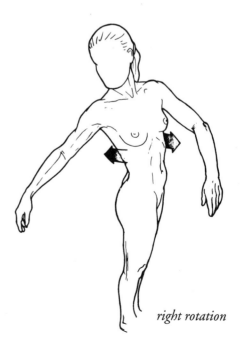

right rotation

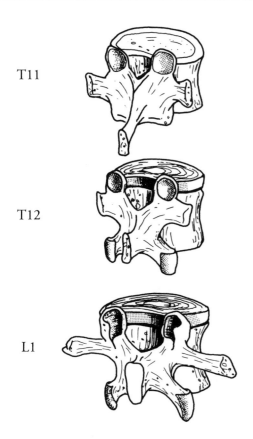

T11

T12

L1

The junction between the thoracic and lumbar spinal regions has some interesting features. T12 resembles T11 in its upper half, but resembles L1 in its lower half, i.e., the spinous process is shortened (which facilitates extension), and the inferior articular facets are large and convex (which restricts rotation).

Because T11 and T12 are both attached to floating ribs, and because of the shape of the spinous process and inferior articular processes on T11, there is good range of motion in all directions at the T11/T12 joint: flexion, extension, sidebending, and rotation.

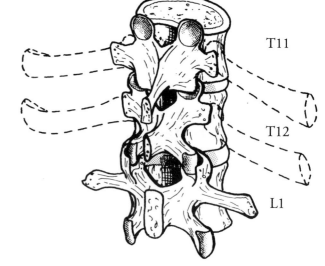

T11

T12

L1

Since this is the first joint above the sacrum which allows good rotation, it is likely to be stressed by excessive rotation.

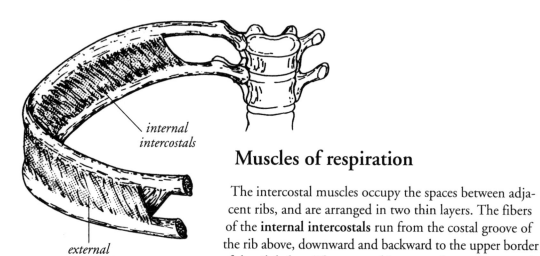

internal intercostals

external intercostals

Muscles of respiration

The intercostal muscles occupy the spaces between adjacent ribs, and are arranged in two thin layers. The fibers of the **internal intercostals** run from the costal groove of the rib above, downward and backward to the upper border of the rib below. The **external intercostals** run downward and *forward* to the rib below, i.e., at right angles to the internal intercostals. The intercostals obviously serve to hold together and preserve the correct shape of the rib cage. There is some controversy about their role in normal respiratory movements. Some electromyographic studies suggest that the externals and internals are involved in inspiration and expiration respectively. Both layers seem to contract during deep, forced inspiration.

Levatores costarum run from the transverse process of a thoracic vertebra to the tubercle of a rib located one or two levels below. They assist in rotation of the spine or elevation of the ribs, depending on which end is fixed.

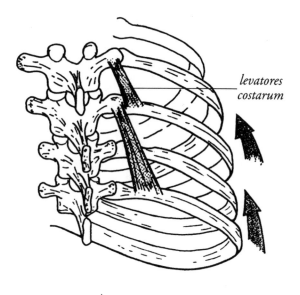

levatores costarum

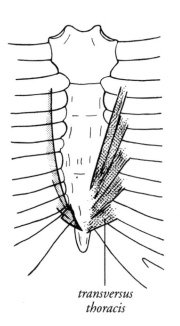

Transversus thoracis originates from the posterior surface of the lower sternum and xiphoid process, and runs superolaterally to insert on the cartilages of ribs 2 through 6. Its contraction lowers those ribs, assisting in expiration.

transversus thoracis

The **diaphragm** is the primary muscle of respiration. It is shaped like a large dome at the inferior end of the thoracic cage, and divides the thorax from the abdomen. The right side of the dome is slightly higher than the left because of the presence of the liver just below it. The level of the top of the dome varies from the 4th intercostal space (during expiration) to the 6th intercostal space (during inspiration).

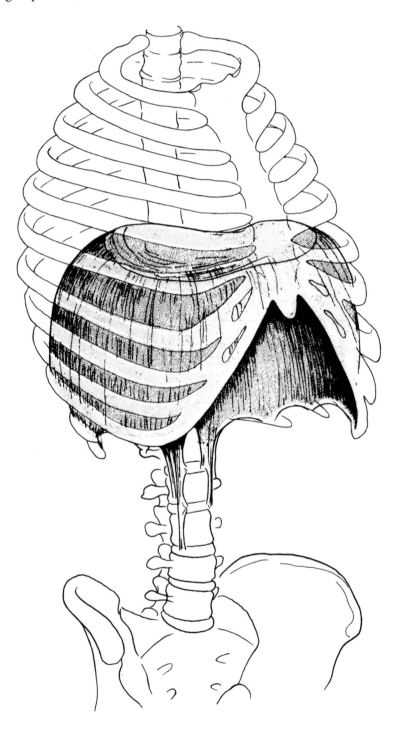

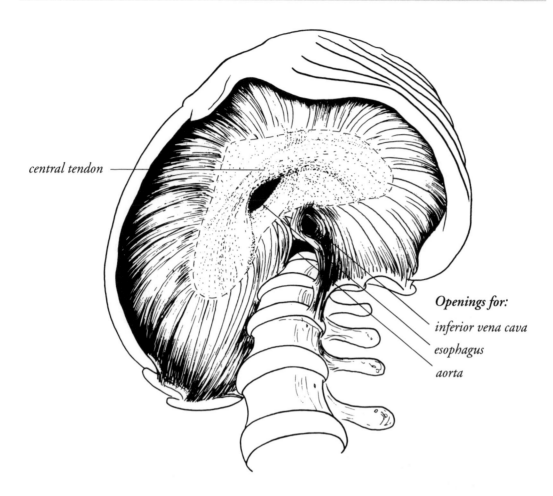

central tendon

Openings for:

inferior vena cava

esophagus

aorta

Unlike any other muscle in the body, the diaphragm inserts not on a bone or other external structure, but on its own central tendon, which is composed of strong interlacing fibrous tissue and shaped somewhat like a three-leafed clover. In the middle and near the vertebral column, the diaphragm is pierced by three large openings for the inferior vena cava, esophagus, and descending aorta.

The muscular portion of the diaphragm has three major origins, all inserting on the central tendon. The **sternal origin** is from the xiphoid process. The **costal origin** is from the deep surfaces of ribs 7-12 and their cartilages; the fibers of attachment interdigitate with those of the transversus abdominis muscle. The **vertebral origin** consists of a "right crus" arising directly from the bodies of L1-L3, a "left crus" arising from the bodies of L1-L2, and five **arcuate ligaments**. The single median arcuate ligament joins the two crura at the midline and arches over the abdominal entrance of the aorta. The paired medial arcuate ligaments extend from the body of L1 to its tranverse processes, arching over the psoas major muscle. The paired lateral arcuate ligaments extend from the transverse processes of L1 to rib 12, arching over the quadratus lumborum muscle.

The inferior surfaces of the lungs are attached to the superior surface of the diaphragm. By contracting, the diaphragm increases the volume of the lungs and causes air to enter the lungs (inspiration). When the diaphragm relaxes, its dome moves superiorly, the volume of the lungs decreases, and air is expelled (expiration).

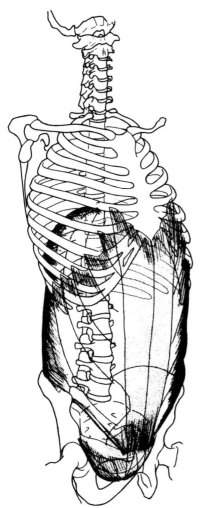

Movements of respiration

The **abdominal cavity** is inferior to the thorax and contains the abdominal organs. It is bounded above by the diaphragm, posteriorly by the lumbar vertebrae, laterally and anteriorly by the abdominal muscles, and inferiorly by the pelvis, and pelvic and urogenital diaphragms.

The abdominal muscles, as well as the diaphragm, play an important role in respiration.

The abdomen can be compared to a liquid-filled, flexible container which can change its shape but not its volume (i.e., is non-compressible). In contrast, the thorax can be compared to a gas-filled container which can change its shape and *is* compressible.

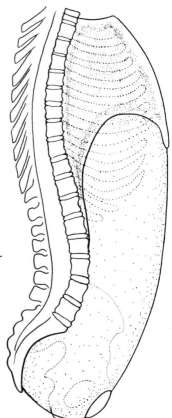

The diaphragm is the interface between these two cavities. In cooperation with the abdominal muscles, it helps control the shape, volume, and pressure of both cavities during numerous activities (breathing, speaking, coughing, defecating, childbirth, hiccups, etc.).

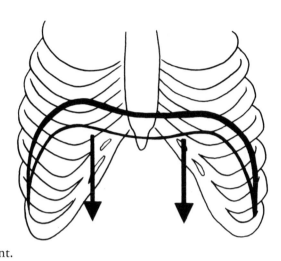

During normal inspiration, contraction of the diaphragm lowers its dome and thereby increases the volume of the thorax, and lungs. The increase in volume results in a decrease of pressure inside the lungs, which causes air to enter from the outside environment.

On the other hand, contraction of the abdominal muscles can cause the abdomen to maintain its shape and resist lowering of the diaphragm. In this case, the central tendon becomes the fixed point and contraction of the fibers of the diaphragm results in elevation of the ribs, because of (i) the superomedial orientation of the fibers, and (ii) lateral pressure from the contents of the abdomen, which are being compressed from above.

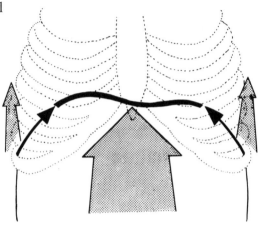

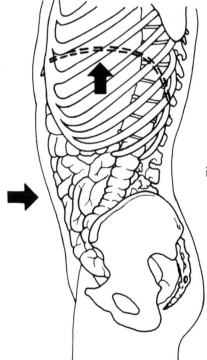

During normal expiration, the diaphragm simply relaxes (i.e., the dome moves upward), and the elastic lung tissue returns to its normal size after being stretched during inspiration. The decrease in lung volume results in an increase in pressure, and air is expelled through the nose or mouth.

In "forced" (or active) expiration, the abdominal muscles contract and press the abdominal organs upward against the diaphragm, further decreasing lung volume and increasing the pressure which drives air out.

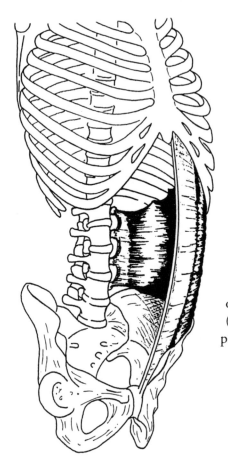

Abdominal muscles

There are four paired muscles in the abdominal wall. They fill all the space (front, sides, and back) between the ribcage and pelvis.

Transversus abdominis is the deepest of the four. It attaches below to the lateral third of the **inguinal ligament** (which runs from the anterior superior iliac spine to the pubic tubercle) and the iliac crest; posteriorly to the thoracolumbar fascia; above to the inner surfaces of ribs 7-12 (where it interdigitates with fibers of the diaphragm); and anteriorly to the **linea alba** (a tough fibrous band stretching from the xiphoid process to the symphysis pubis).

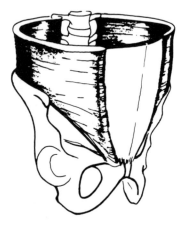

The fibers of the transversus are essentially horizontal. They terminate anteriorly in a broad aponeurosis.

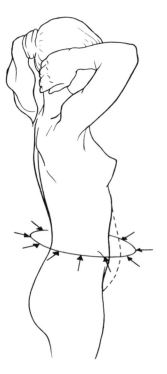

Contraction of these circular fibers reduces the diameter of the abdomen, i.e., "pulls in the belly" or increases lordosis of the lumbar spine. An easy way to feel the action of the transversus is to wrap your hands around the sides of your abdomen and cough.

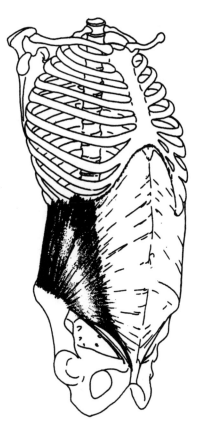

The **internal oblique** lies between the transversus and external oblique. It is attached below to the lateral half of the inguinal ligament and the iliac crest; posteriorly to the thoracolumbar fascia; above, to the lower ribs; and anteriorly to a very broad aponeurosis.

The fibers run in various directions depending on their location, but the "average" direction is anterosuperior. Unilateral contraction of the internal oblique results in sidebending or ipsilateral rotation of the spine and ribcage.

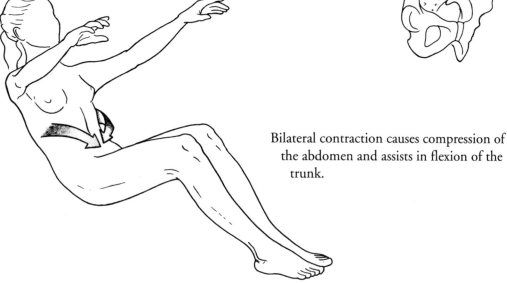

Bilateral contraction causes compression of the abdomen and assists in flexion of the trunk.

The **external oblique** is external to the two muscles described above. It is attached above to the outer surfaces of ribs 5-12 (where its fibers intertwine with those of the serratus anterior and latissimus dorsi). In front and below, it forms a broad aponeurosis ending at (and contributing to) the linea alba and inguinal ligament.

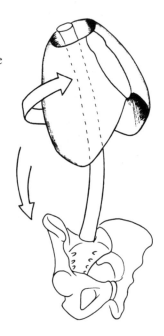

The "average" direction of the fibers is anteroinferior, i.e., perpendicular to those of the internal oblique. Unilateral contraction of the external oblique results in side-bending and contralateral rotation of the spine and ribcage.

Again, bilateral contraction causes compression of the abdomen and assists in flexion of the trunk.

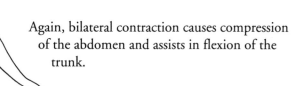

The oblique muscles work synergetically in rotation of the trunk. For instance, rotation of the trunk to the right (combined with flexion) involves simultaneous contraction of the right internal oblique and left external oblique.

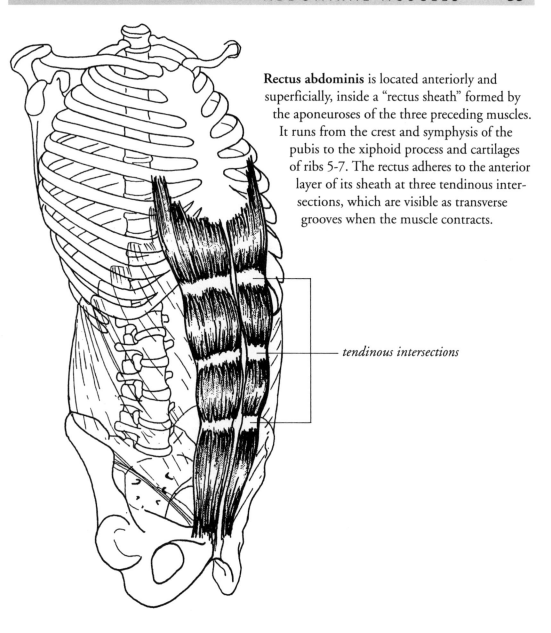

Rectus abdominis is located anteriorly and superficially, inside a "rectus sheath" formed by the aponeuroses of the three preceding muscles. It runs from the crest and symphysis of the pubis to the xiphoid process and cartilages of ribs 5-7. The rectus adheres to the anterior layer of its sheath at three tendinous intersections, which are visible as transverse grooves when the muscle contracts.

tendinous intersections

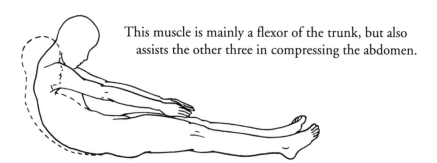

This muscle is mainly a flexor of the trunk, but also assists the other three in compressing the abdomen.

There are three important muscles of the lower abdomen.

Obturator internus forms the lateral walls of the pelvis. It originates from the inner aspect of the pubis and ischium, and obturator membrane (which covers the obturator foramen), and inserts on the greater trochanter of the femur.

The **pelvic diaphragm** forms the floor of the pelvis, and consists of two paired muscles. **Levator ani** originates from fascia covering the obturator internus, along a line extending from the posterior body of the pubis to the ischial spine. The fibers from the left and right sides of this muscle run inferomedially and meet each other at the midline. Some posterior fibers insert on the coccyx. Levator ani contains an opening for the anal canal. **Coccygeus** is posterior to levator ani, running from the ischial spine to the coccyx and lower sacrum. Both these muscles support the weight of the pelvic organs, and are involved in the movements of defecation. In ancestral mammals, coccygeus was responsible for wagging the tail.

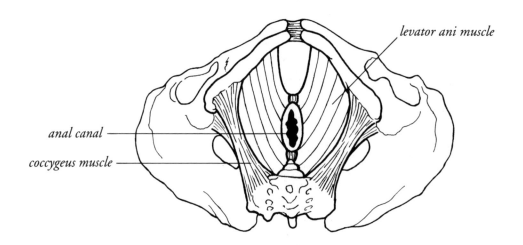

Inferior to the pelvic diaphragm, and overlapping it anteriorly, is the **urogenital diaphragm**, whose fibers run transversely from the rami of the pubis and ischium on one side to those on the other. The urogenital diaphragm consists mainly of the **sphincter urethrae** muscles. It contains openings for the urethra in both sexes, and the vagina in females.

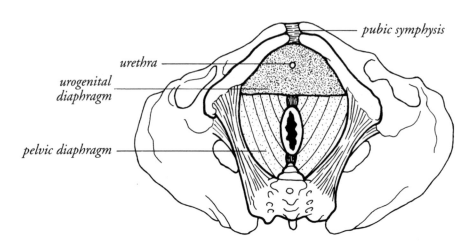

CHAPTER THREE

The Shoulder

··

The shoulder is the area where the arm is attached to the thorax. In contrast to the hip, it involves more than one joint. This complex structure has two important but somewhat contradictory functions:

- It must be very flexible, to allow the hand and arm the huge range of motion which they require.

- It must provide a strong, stable fixed point for certain actions (lifting a heavy object, pushing against resistance, etc.)

The primary joint of the shoulder is the **glenohumeral joint** between the head of the humerus and glenoid cavity of the scapula. However, the scapula itself is an extremely mobile bone, and is connected to the axial skeleton (i.e., sternum) only by the long, thin clavicle. Thus, two other joints are involved in movements of the shoulder: the **acromioclavicular joint** between the distal clavicle and acromion process of the scapula, and the **sternoclavicular joint** between the medial clavicle and manubrium of the sternum.

Landmarks

Some visible and palpable landmarks of the shoulder are shown below.

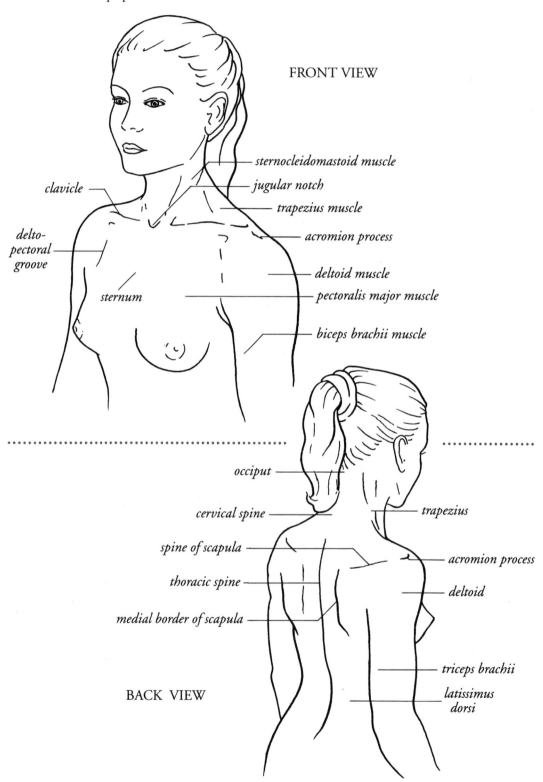

FRONT VIEW

sternocleidomastoid muscle

jugular notch

trapezius muscle

acromion process

clavicle

delto-pectoral groove

deltoid muscle

pectoralis major muscle

sternum

biceps brachii muscle

occiput

cervical spine

trapezius

spine of scapula

acromion process

thoracic spine

deltoid

medial border of scapula

triceps brachii

latissimus dorsi

BACK VIEW

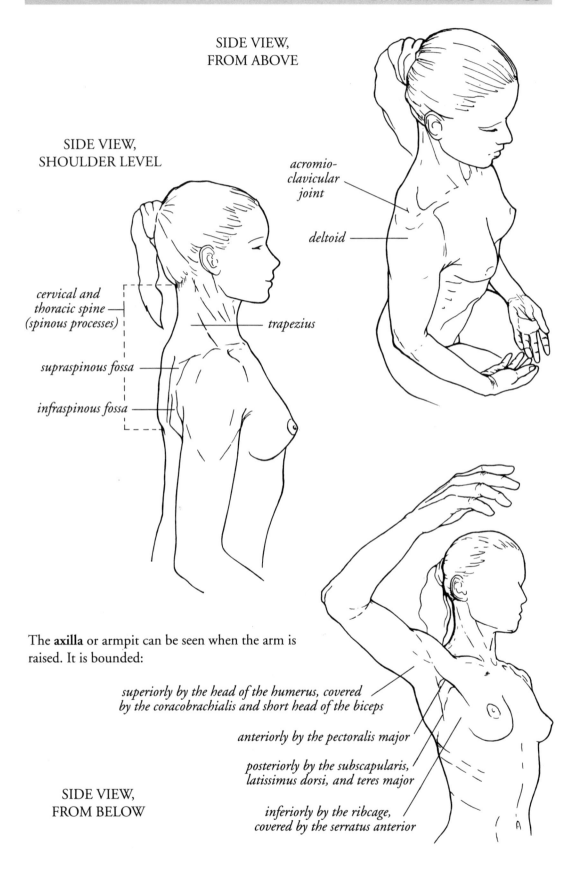

SIDE VIEW,
FROM ABOVE

SIDE VIEW,
SHOULDER LEVEL

acromio-
clavicular
joint

deltoid

cervical and
thoracic spine
(spinous processes)

trapezius

supraspinous fossa

infraspinous fossa

The **axilla** or armpit can be seen when the arm is
raised. It is bounded:

*superiorly by the head of the humerus, covered
by the coracobrachialis and short head of the biceps*

anteriorly by the pectoralis major

*posteriorly by the subscapularis,
latissimus dorsi, and teres major*

*inferiorly by the ribcage,
covered by the serratus anterior*

SIDE VIEW,
FROM BELOW

Movements

The scapula lies very close to the posterior ribcage (at the level of ribs 2 through 7) but does not articulate with it. Rather it "floats" behind it, suspended in a network of muscles and ligaments. It can move in many directions:

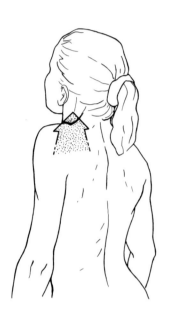
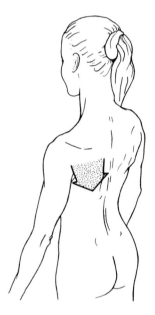
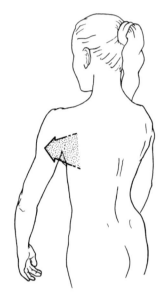

superiorly (elevation)

inferiorly (depression)

laterally (protraction or abduction)

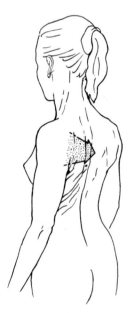
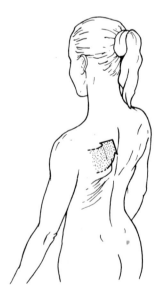
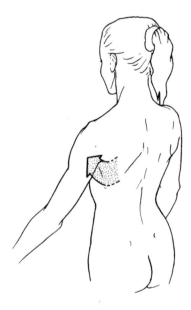

medially (retraction or adduction)

rotation (inferior angle moves towards midline)

rotation (inferior angle moves away from midline)

The great mobility of the scapula combined with actions of the arm muscles allows the arm to move in many ways:

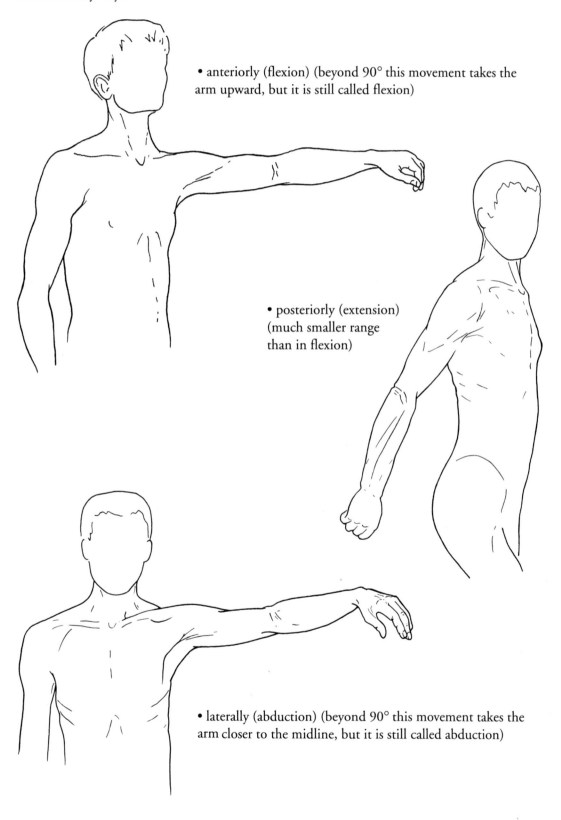

• anteriorly (flexion) (beyond 90° this movement takes the arm upward, but it is still called flexion)

• posteriorly (extension) (much smaller range than in flexion)

• laterally (abduction) (beyond 90° this movement takes the arm closer to the midline, but it is still called abduction)

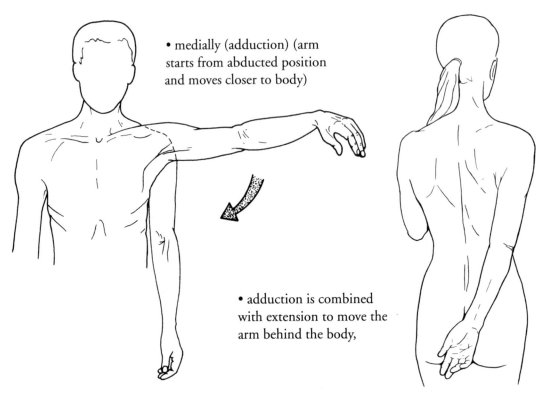

• medially (adduction) (arm starts from abducted position and moves closer to body)

• adduction is combined with extension to move the arm behind the body,

...or with flexion to move the arm in front of the body (below)

• rotation of the humerus (on its axis) is best visualized with the elbow bent, to avoid confusion with pronation and supination of the forearm. In medial rotation, an imaginary point on the front of the humerus moves closer to the midline (and, in this picture, the forearm moves anteriorly). In lateral rotation, this point moves away from the midline (forearm moves posteriorly).

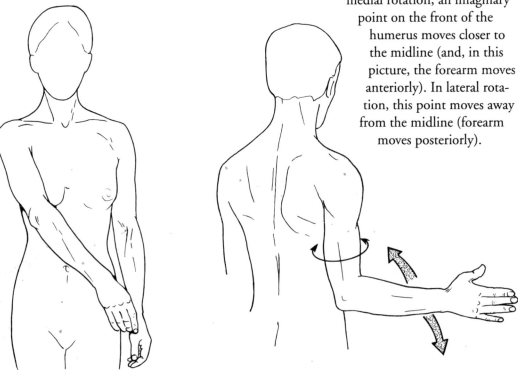

At their extremes, arm movements affect the thoracic spine and ribcage.

• Flexion of the arms is associated with extension of the spine and "opening" of the anterior ribcage (i.e., ribs move apart from each other).

• Extension of the arms causes flexion of the spine and "closing" of the ribcage.

• Adduction is associated with ipsilateral side-bending and closing of the ipsilateral hemithorax...

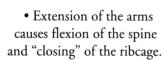

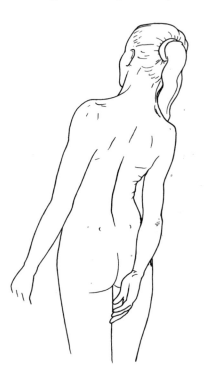

...while abduction causes contralateral sidebending and opening of the ipsilateral hemithorax.

Medial or lateral rotation of the humerus is associated with similar rotation of the spine.

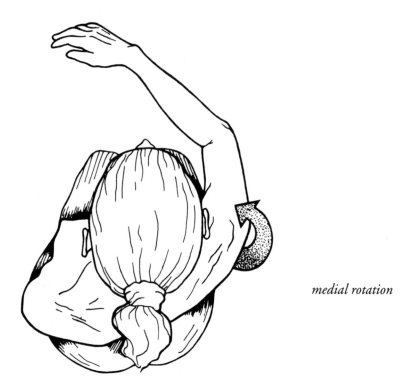

medial rotation

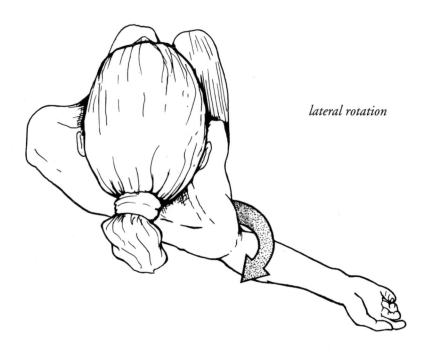

lateral rotation

Clavicle

The shoulder girdle provides a mobile base for movements of the arms. It consists of the scapulae posteriorly, and the clavicles and manubrium anteriorly.

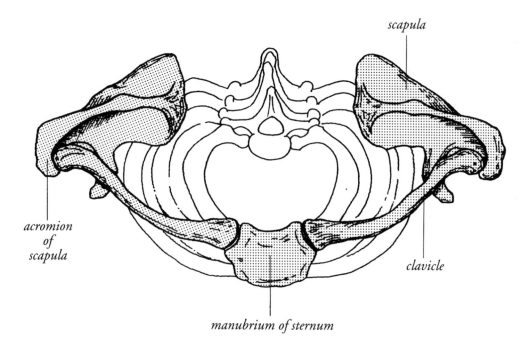

scapula

acromion of scapula

clavicle

manubrium of sternum

The **clavicle** (collarbone) is a flattened, elongated bone with two bends in it, so that it is roughly S-shaped when viewed from above or below.

It articulates medially with the manubrium, and laterally with the acromion. Because of its fragility and role of transmitting forces from the arms to the thorax (e.g., in falling on one's outstretched arm), it is the most frequently broken bone in the body.

Near the medial and lateral ends of the clavicle are the costoclavicular and coracoclavicular ligaments which anchor it, respectively, to the first costal cartilage and the coracoid process of the scapula. If you examine a clavicle, you will see roughened areas for attachment of these ligaments.

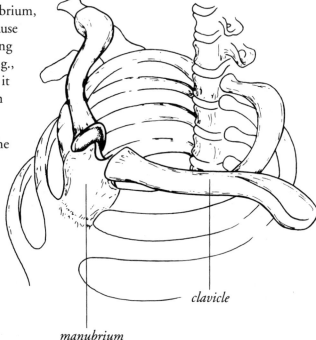

clavicle

manubrium

Sternoclavicular joint

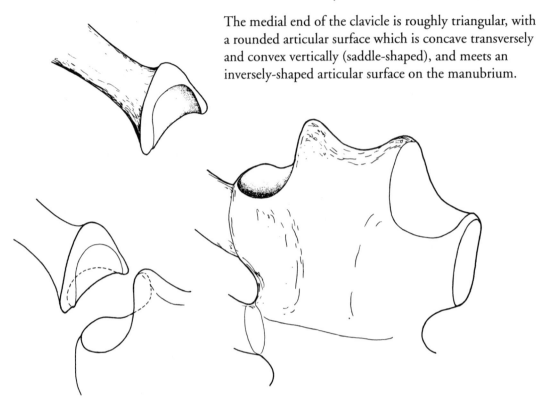

The medial end of the clavicle is roughly triangular, with a rounded articular surface which is concave transversely and convex vertically (saddle-shaped), and meets an inversely-shaped articular surface on the manubrium.

  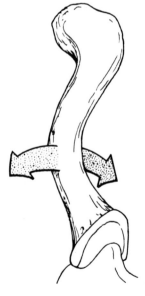

This saddle joint allows flexion and extension,

elevation and depression,

and limited rotation of the clavicle.

These movements generally occur secondarily in association with movements of the scapula.

Scapula

The scapula, or shoulderblade, is is a flat triangular bone with three borders and three angles, as shown.

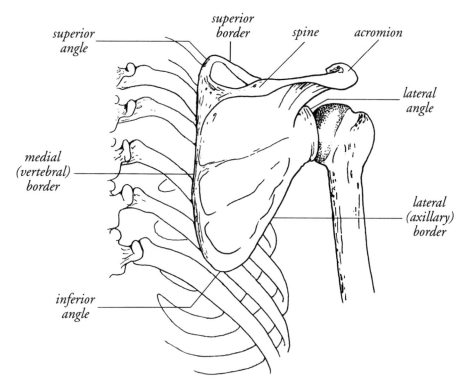

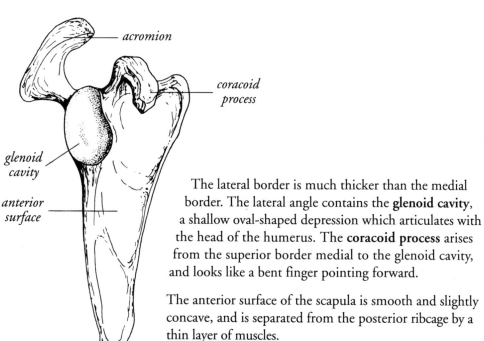

The lateral border is much thicker than the medial border. The lateral angle contains the **glenoid cavity**, a shallow oval-shaped depression which articulates with the head of the humerus. The **coracoid process** arises from the superior border medial to the glenoid cavity, and looks like a bent finger pointing forward.

The anterior surface of the scapula is smooth and slightly concave, and is separated from the posterior ribcage by a thin layer of muscles.

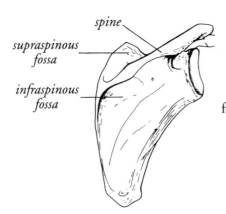

spine

supraspinous
fossa

infraspinous
fossa

The posterior surface is much rougher, since many muscles attach to it. The **spine** is a strong, sharp ridge running diagonally near the superior border. It is easily palpated and seen externally. There are depressions above and below it called the supra-spinous and infraspinous fossae.

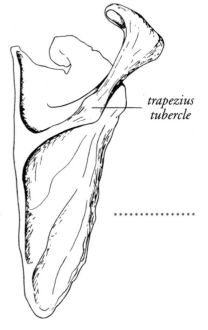

trapezius
tubercle

The lateral end of the spine is enlarged and flattened to form the acromion process. Note that the acromion is oriented perpendicular to the rest of the spine. There is an enlarged tubercle located more medially on the spine for attachment of the trapezius.

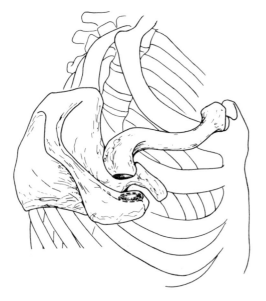

Acromioclavicular joint

The acromion and the lateral end of the clavicle each have a small oval surface where they articulate (left). Sometimes this joint includes a meniscus (fibrous disk). The shape of the articular surfaces allows some gliding movement, as well as opening and closing of the angle formed by the two bones.

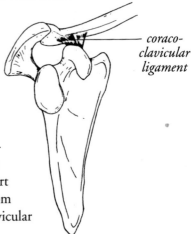

coraco-
clavicular
ligament

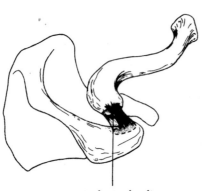

The capsule is loose, and thickened on its superior aspect to form an acromioclavicular ligament (left). The main support for this joint comes from the extrinsic coracoclavicular ligament (right).

acromioclavicular ligament

Movements of the scapula

These movements were shown in external view on page 102. We will now look at the bones themselves.

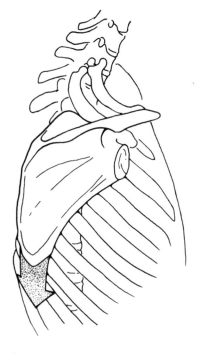

In **elevation**, the scapula moves upward and away from the ribcage.

In **depression**, it moves downward and fits more snugly against the ribcage.

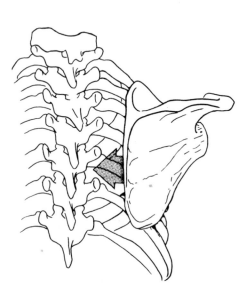

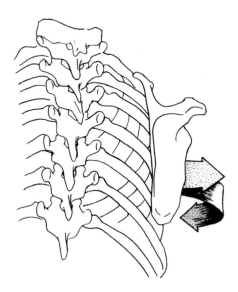

In **retraction (adduction)**, the medial border moves closer to the vertebral column, and the lateral angle moves posteriorly.

In **protraction (abduction)**, the medial border moves away from the vertebral column, and the lateral angle moves anteriorly.

To understand the two types of **rotation**, it is helpful to imagine a "pivot" (fixed point) for the scapula, passing through the middle of the infraspinous fossa.

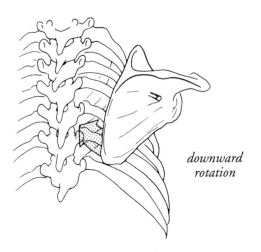

downward rotation

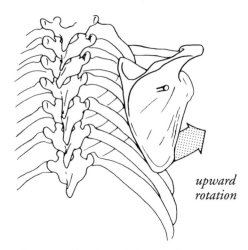

upward rotation

In one type of rotation, the inferior angle moves superomedially, while the lateral angle moves inferolaterally. We call this "downward" rotation because the glenoid cavity is moving downward.

Alternatively, the inferior angle can move superolaterally, while the superior angle moves inferomedially. We call this "upward" rotation, again using the glenoid cavity as our reference point.

Mobility of the arm is greatly increased by such movements of the scapula. For example, consider abduction of the arm with the scapula fixed (left) vs. the scapula rotated (below).

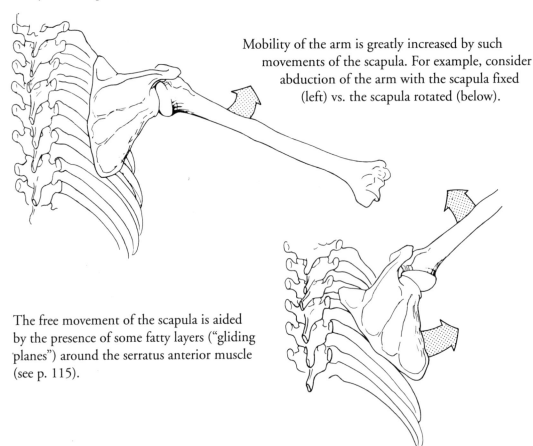

The free movement of the scapula is aided by the presence of some fatty layers ("gliding planes") around the serratus anterior muscle (see p. 115).

Humerus

This is the long bone of the upper arm. Landmarks of the proximal end include the medial **head** (which articulates with the glenoid cavity of the scapula), the lateral **greater tubercle** and anterior **lesser tubercle** for muscle attachment, the **bicipital (intertubercular) groove** running between the two tubercles, and the **anatomical neck** just below the head.

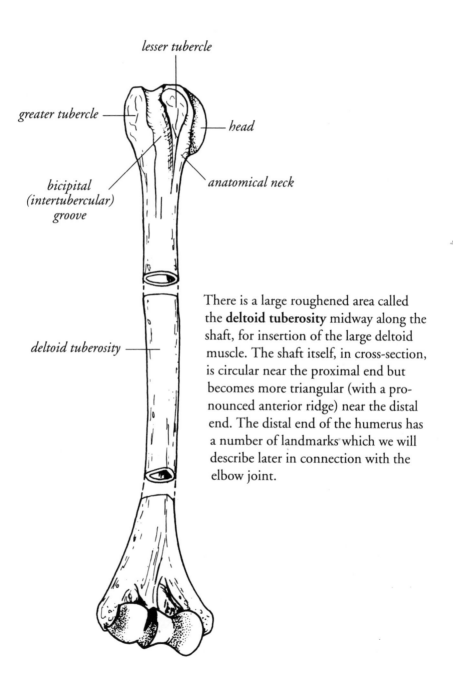

lesser tubercle

greater tubercle

head

bicipital (intertubercular) groove

anatomical neck

deltoid tuberosity

There is a large roughened area called the **deltoid tuberosity** midway along the shaft, for insertion of the large deltoid muscle. The shaft itself, in cross-section, is circular near the proximal end but becomes more triangular (with a pronounced anterior ridge) near the distal end. The distal end of the humerus has a number of landmarks which we will describe later in connection with the elbow joint.

Glenohumeral joint

This is the primary joint of the shoulder. The head of the humerus represents about ⅔ of a sphere. Its orientation is mostly medial, but slightly posterosuperior. The surface area of the head is two or three times larger than that of the glenoid cavity, and the cavity is fairly shallow. Thus, there is not a "snug fit" between the two bones, and the shoulder joint (compared to the hip joint) is extremely mobile but much less stable. The orientation of the glenoid cavity is mostly lateral, but slightly anterosuperior.

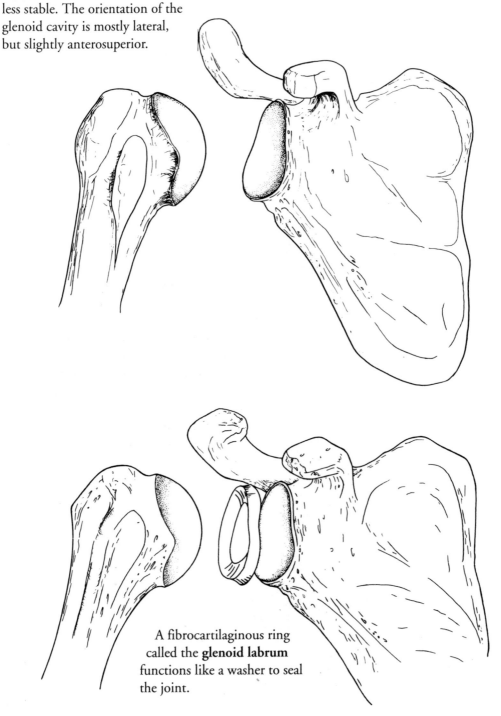

A fibrocartilaginous ring called the **glenoid labrum** functions like a washer to seal the joint.

The capsule of the glenohumeral joint attaches to the scapula, on the outer rim of the glenoid cavity. Superiorly, it goes all the way up to the coracoid process, and encircles the origin of the long head of the biceps at its origin. On the humerus, it attaches to the anatomical neck. The capsule is quite loose and has many folds, especially inferiorly where it is weakest. The strongest ligament reinforcing the capsule is the superior **coracohumeral ligament** which runs from the border of the coracoid process to the greater tubercle. Anteriorly, there are three **glenohumeral ligaments**, which run from the border of the glenoid cavity to the lesser tubercle and anatomical neck of the humerus.

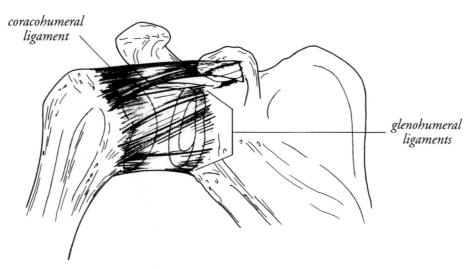

coracohumeral ligament

glenohumeral ligaments

The capsule is also reinforced by blending with the tendons of the **rotator cuff** muscles (see pages 120-122). The capsule is weakest anteroinferiorly where it has no support from these muscles, and no ligaments. The most common type of dislocation of the glenohumeral joint involves anteromedial movement of the humeral head. The capsular ligaments may be strained or torn during such dislocations (right).

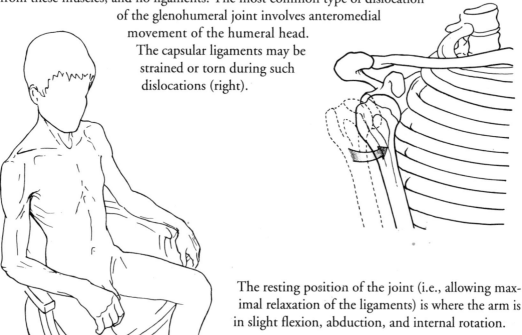

The resting position of the joint (i.e., allowing maximal relaxation of the ligaments) is where the arm is in slight flexion, abduction, and internal rotation.

Shoulder muscles

There are many muscles which move the scapula, clavicle, and humerus. We will briefly describe the locations and actions of the most important ones.

Serratus anterior is a broad, thin muscle covering the lateral ribcage. It originates from the upper eight or nine ribs, and inserts along the entire medial border of the scapula. It helps hold (fix) the scapula in place, and also functions in abduction and rotation. Its fibers are contracted and visible when the arm is pushing against some resistance (below). When the scapula is fixed, the inferior fibers of serratus assist in inspiration by helping to elevate the ribs.

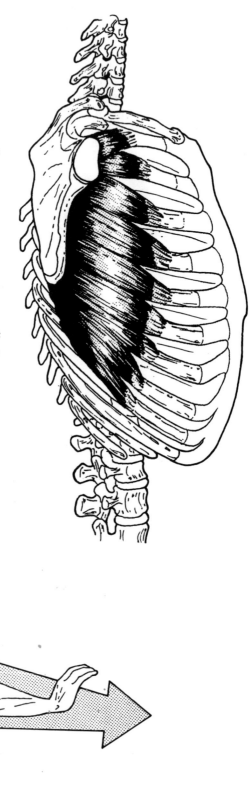

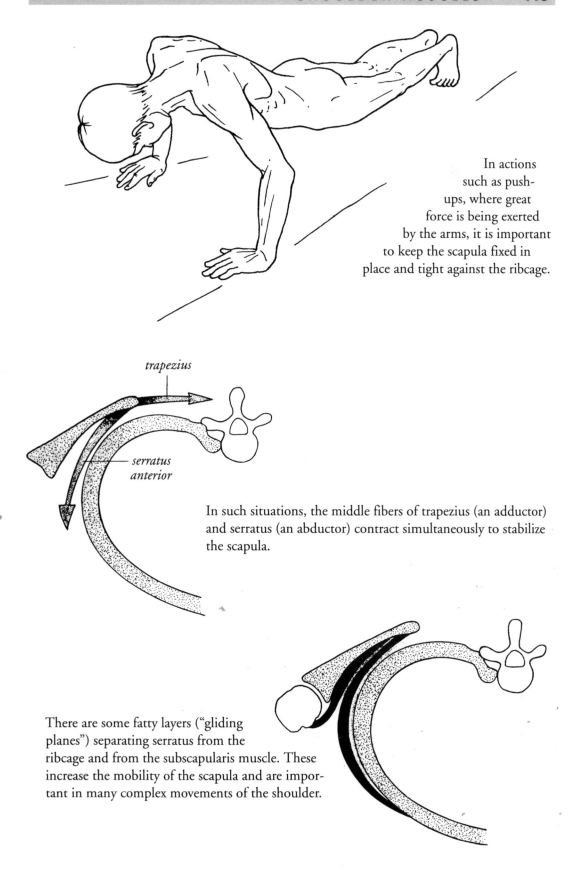

In actions such as push-ups, where great force is being exerted by the arms, it is important to keep the scapula fixed in place and tight against the ribcage.

trapezius

serratus anterior

In such situations, the middle fibers of trapezius (an adductor) and serratus (an abductor) contract simultaneously to stabilize the scapula.

There are some fatty layers ("gliding planes") separating serratus from the ribcage and from the subscapularis muscle. These increase the mobility of the scapula and are important in many complex movements of the shoulder.

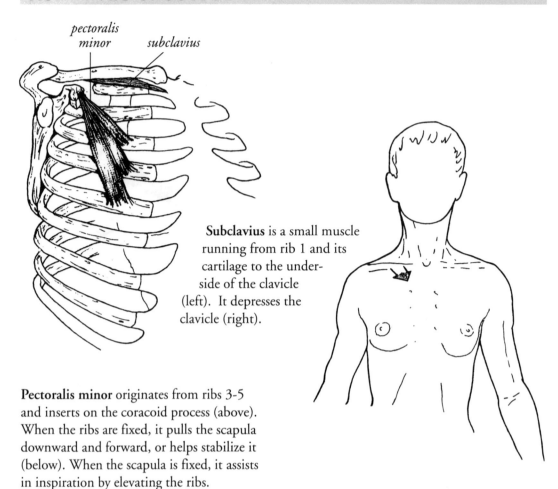

pectoralis minor   *subclavius*

Subclavius is a small muscle running from rib 1 and its cartilage to the underside of the clavicle (left). It depresses the clavicle (right).

Pectoralis minor originates from ribs 3-5 and inserts on the coracoid process (above). When the ribs are fixed, it pulls the scapula downward and forward, or helps stabilize it (below). When the scapula is fixed, it assists in inspiration by elevating the ribs.

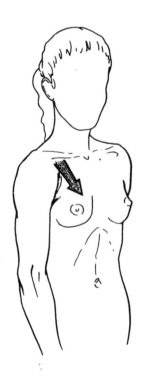

Sternocleidomastoid (SCM) was described in Chapter 2. It acts primarily on the head and cervical spine. If the head is fixed, however, SCM can elevate the area where the clavicle and sternum meet, and thereby assist in inspiration.

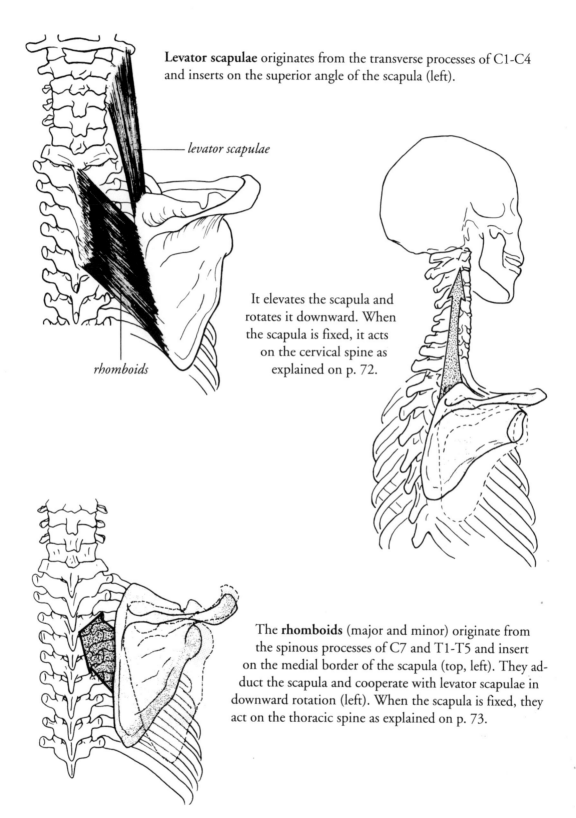

Levator scapulae originates from the transverse processes of C1-C4 and inserts on the superior angle of the scapula (left).

levator scapulae

rhomboids

It elevates the scapula and rotates it downward. When the scapula is fixed, it acts on the cervical spine as explained on p. 72.

The **rhomboids** (major and minor) originate from the spinous processes of C7 and T1-T5 and insert on the medial border of the scapula (top, left). They adduct the scapula and cooperate with levator scapulae in downward rotation (left). When the scapula is fixed, they act on the thoracic spine as explained on p. 73.

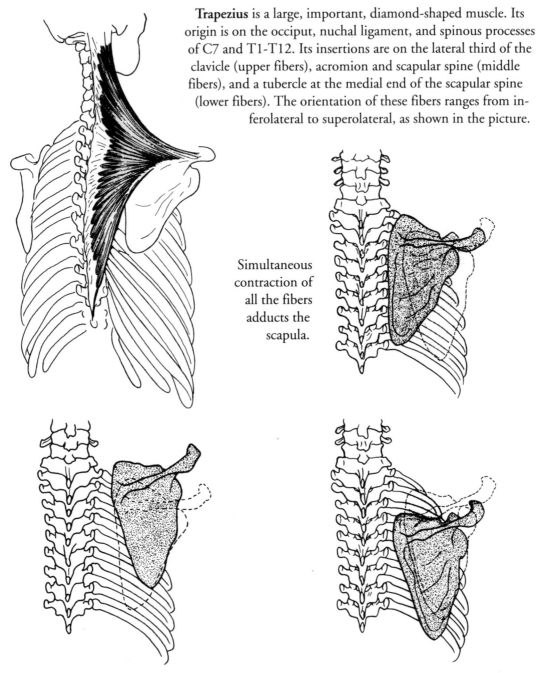

Trapezius is a large, important, diamond-shaped muscle. Its origin is on the occiput, nuchal ligament, and spinous processes of C7 and T1-T12. Its insertions are on the lateral third of the clavicle (upper fibers), acromion and scapular spine (middle fibers), and a tubercle at the medial end of the scapular spine (lower fibers). The orientation of these fibers ranges from inferolateral to superolateral, as shown in the picture.

Simultaneous contraction of all the fibers adducts the scapula.

The upper fibers by themselves act in elevation and upward rotation of the scapula, and elevation of the clavicle.

The lower fibers by themselves act in depression and upward rotation of the scapula.

The upper fibers are frequently over-solicited in work (e.g., at a computer or typewriter) involving prolonged suspension of the arms, with resulting symptoms of sore neck, muscle stiffness or spasm, headache, etc. When force needs to be exerted or absorbed by the arm, the middle fibers of trapezius (adductor) act in synergy with serratus anterior (abductor) to stabilize the scapula (see p. 115).

Muscles involved in specific movements of the scapula

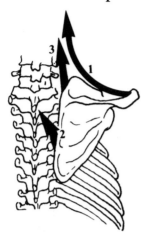

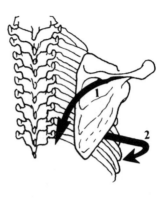

elevation: (1) upper trapezius;
(2) rhomboids; (3) levator scapulae

depression: (1) lower trapezius;
(2) lower serratus anterior

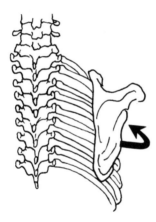

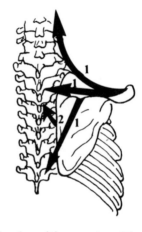

abduction: serratus anterior

adduction: (1) trapezius; (2) rhomboids

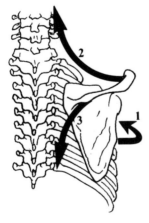

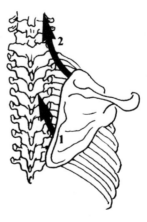

upward rotation: (1) serratus anterior;
(2) upper trapezius; (3) lower trapezius

downward rotation: (1) rhomboids;
(2) levator scapulae

Arm muscles

Anatomists use the word "arm" to refer to the humerus. "Forearm" is used to refer to the radius and ulna. The muscles described in this section all have their bodies located near the scapula or upper thorax, but they insert and act on the humerus.

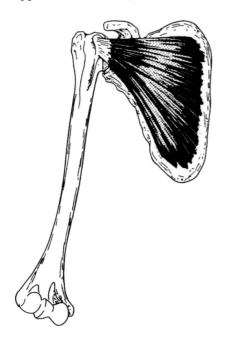

Rotator cuff muscles

Subscapularis originates from the anterior surface of the scapula and inserts on the lesser tubercle of the humerus. This is one of the rotator cuff muscles which reinforce the capsule of the glenohumeral joint (see p. 113).

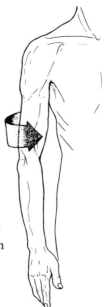

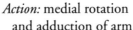

Action: medial rotation and adduction of arm

Supraspinatus originates from the supraspinous fossa on the posterior scapula. Its tendon passes under the acromioclavicular joint and the ligament which connects the coracoid process to the acromion, and inserts on the highest point on the greater tubercle (below left). Deltoid is usually the primary abductor, but supraspinatus can lift the arm by itself even if deltoid is paralyzed. Supraspinatus is an important rotator cuff muscle. There is a large bursa (closed sac of synovial fluid) surrounding its tendon and separating it from the inferior surface of the acromion and deltoid. This bursa acts as an auxiliary component of the glenohumeral joint. Adhesions here can restrict mobility of the shoulder.

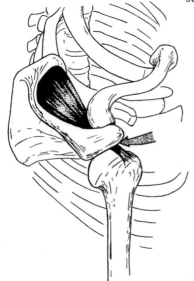

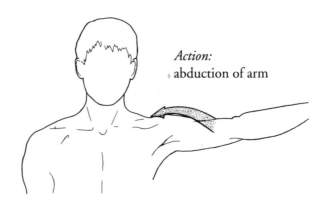

Action: abduction of arm

Infraspinatus originates from the infraspinous fossa and inserts on the greater tubercle at a point posteroinferior to the insertion of supraspinatus.

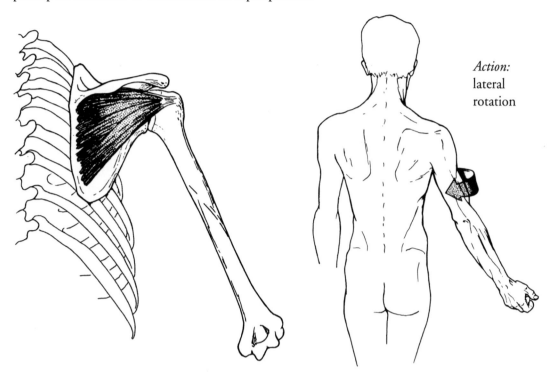

Action:
lateral
rotation

Teres minor originates from the lateral border of the scapula (posterior surface) and inserts on the greater tubercle below the insertion of infraspinatus. This is the fourth and last rotator cuff muscle.

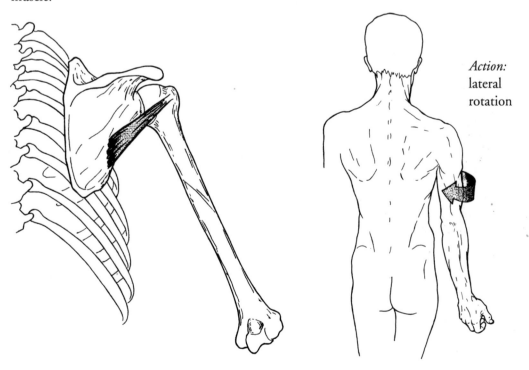

Action:
lateral
rotation

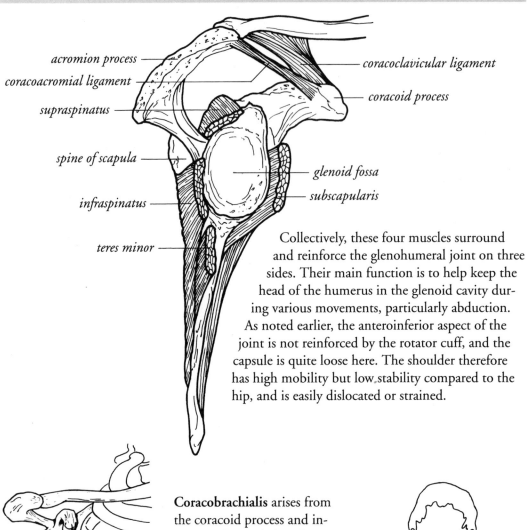

acromion process
coracoacromial ligament
supraspinatus
spine of scapula
infraspinatus
teres minor
coracoclavicular ligament
coracoid process
glenoid fossa
subscapularis

Collectively, these four muscles surround and reinforce the glenohumeral joint on three sides. Their main function is to help keep the head of the humerus in the glenoid cavity during various movements, particularly abduction. As noted earlier, the anteroinferior aspect of the joint is not reinforced by the rotator cuff, and the capsule is quite loose here. The shoulder therefore has high mobility but low stability compared to the hip, and is easily dislocated or strained.

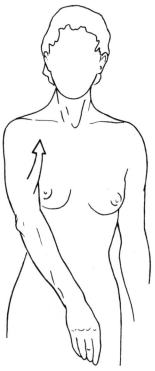

Coracobrachialis arises from the coracoid process and inserts on the medial surface of the humeral shaft, near the middle.

Action: flexes and adducts the arm

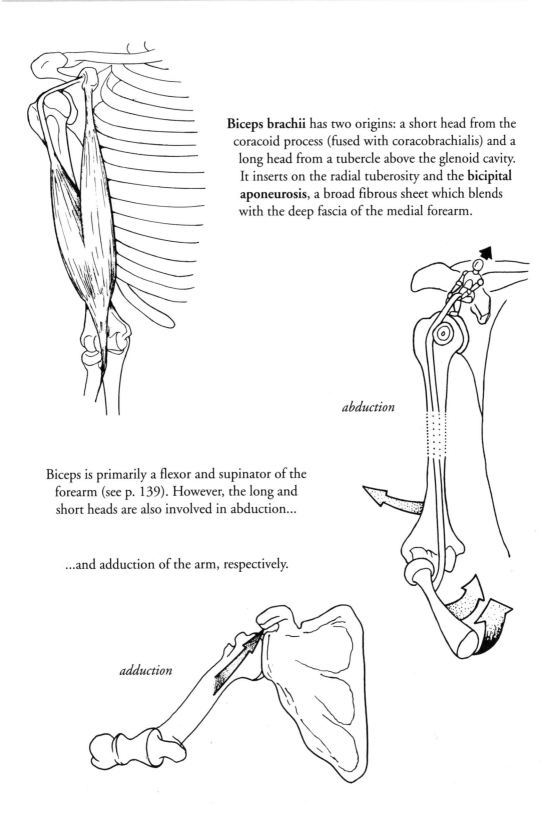

Biceps brachii has two origins: a short head from the coracoid process (fused with coracobrachialis) and a long head from a tubercle above the glenoid cavity. It inserts on the radial tuberosity and the **bicipital aponeurosis**, a broad fibrous sheet which blends with the deep fascia of the medial forearm.

Biceps is primarily a flexor and supinator of the forearm (see p. 139). However, the long and short heads are also involved in abduction...

...and adduction of the arm, respectively.

abduction

adduction

Triceps brachii is primarily a forearm extensor (see p. 140). However, one of its three origins (the long head) is from the scapula, and this part of the muscle assists in adduction of the arm.

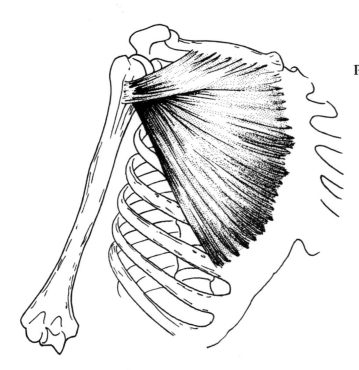

Pectoralis major has a clavicular head from the anterior, medial clavicle, and a sternocostal head from the sternum and costal cartilages 1-6. This large muscle inserts on the lateral aspect of the bicipital groove. The tendon is twisted such that the fibers from the clavicular head insert below those from the sternocostal head.

Contraction of both heads produces adduction and medial rotation of the arm (below left). The clavicular head flexes the extended shoulder, whereas the sternocostal head extends the flexed shoulder. In pull-ups (i.e., grasping a bar with both hands and raising your chin to the bar), pectoralis major pulls the thorax toward the "fixed" arm. The fibers of this muscle can be stretched in the position shown (below right).

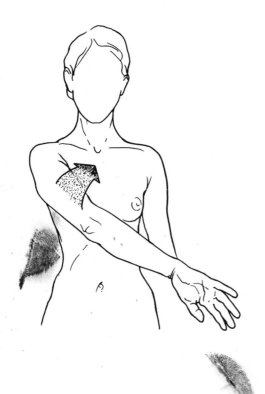

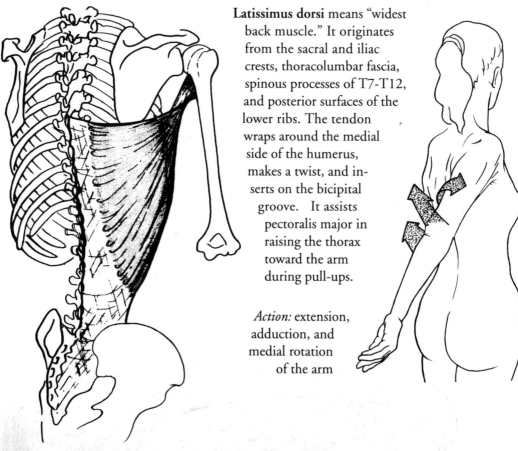

Latissimus dorsi means "widest back muscle." It originates from the sacral and iliac crests, thoracolumbar fascia, spinous processes of T7-T12, and posterior surfaces of the lower ribs. The tendon wraps around the medial side of the humerus, makes a twist, and inserts on the bicipital groove. It assists pectoralis major in raising the thorax toward the arm during pull-ups.

Action: extension, adduction, and medial rotation of the arm

Teres major originates from the posterior surface of the inferior angle of the scapula, and inserts on the medial aspect of the bicipital groove, next to latissimus dorsi. Electromyographic studies indicate that it is active in arm swinging during walking. Note that despite the similarity in name and location, teres major and teres minor have quite distinct functions (and different innervation), and should not be confused with each other.

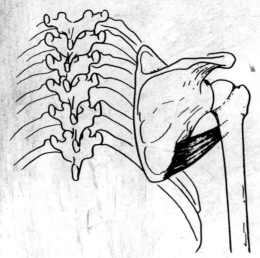

Action: same as latissimus dorsi

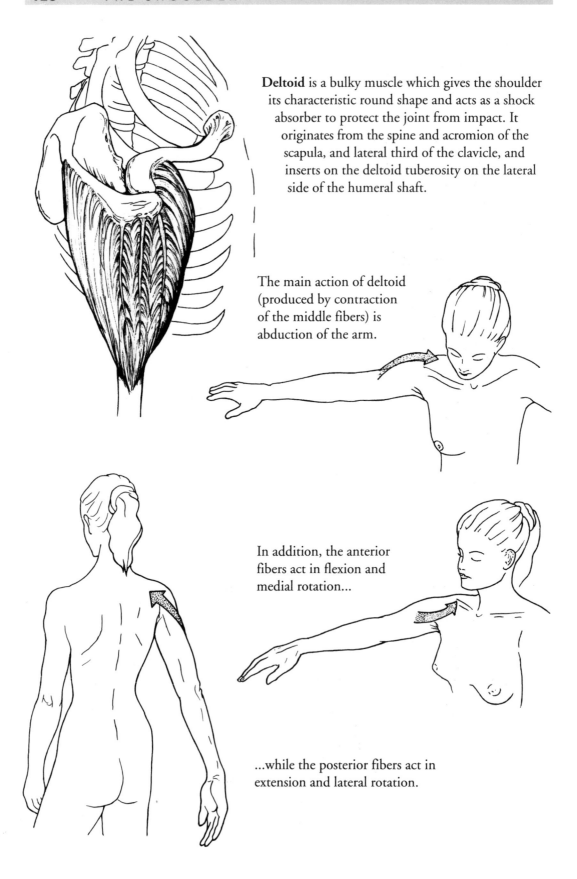

Deltoid is a bulky muscle which gives the shoulder its characteristic round shape and acts as a shock absorber to protect the joint from impact. It originates from the spine and acromion of the scapula, and lateral third of the clavicle, and inserts on the deltoid tuberosity on the lateral side of the humeral shaft.

The main action of deltoid (produced by contraction of the middle fibers) is abduction of the arm.

In addition, the anterior fibers act in flexion and medial rotation...

...while the posterior fibers act in extension and lateral rotation.

Muscles involved in specific movements of the arm

flexion:
(1) anterior deltoid;
(2) pectoralis major;
(3) coracobrachialis

Accessory: biceps brachii, subscapularis

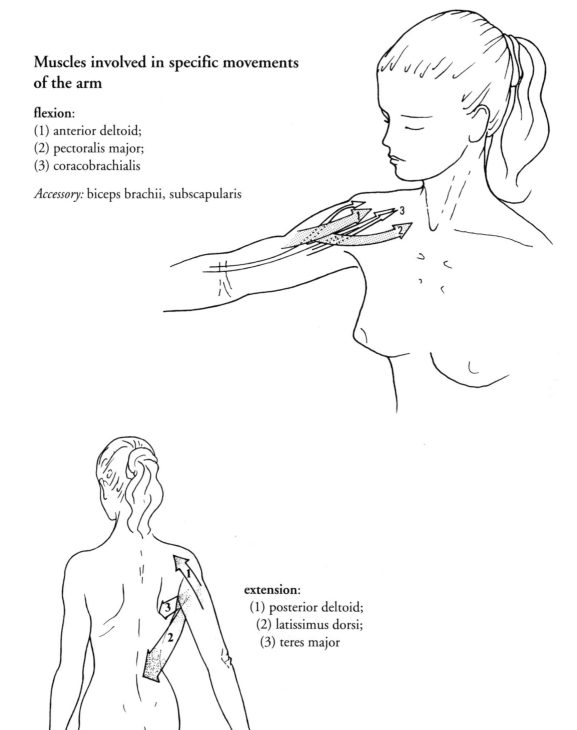

extension:
(1) posterior deltoid;
(2) latissimus dorsi;
(3) teres major

abduction:
(1) deltoid;
(2) supraspinatus

Accessory: infraspinatus, long head of biceps

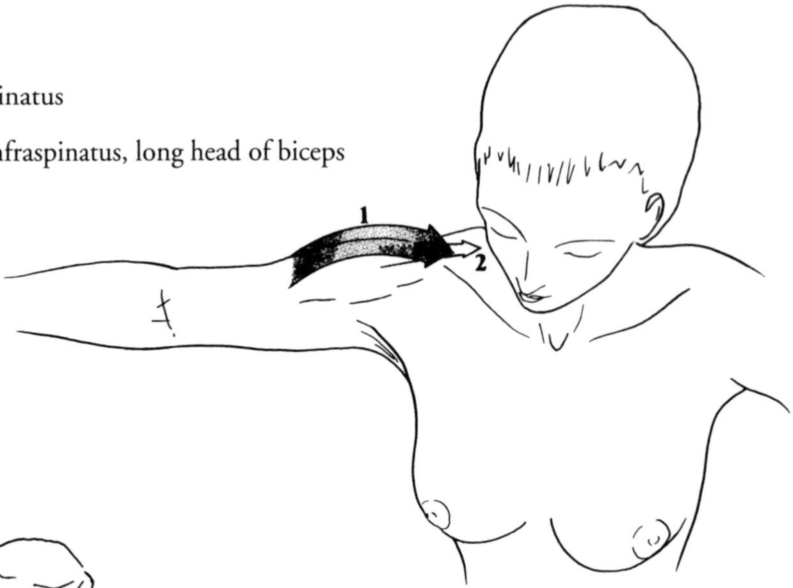

adduction:
(1) latissimus dorsi;
(2) pectoralis major;
 (3) teres major

Accessory: teres minor, short head
of biceps, coracobrachialis

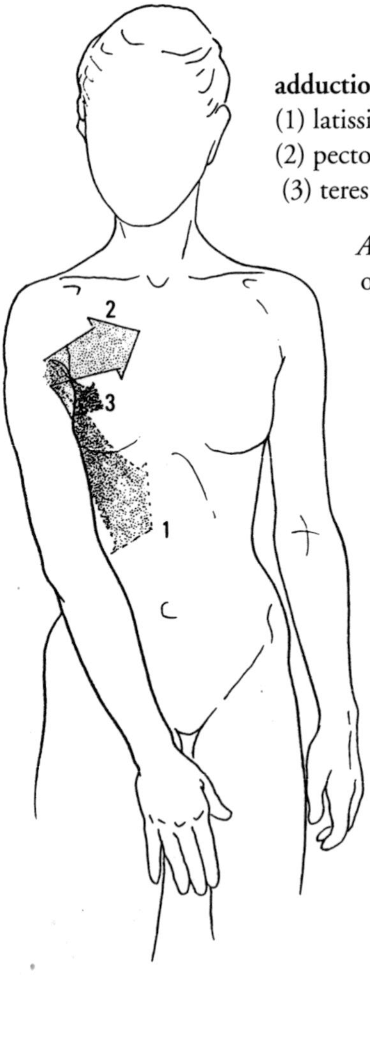

lateral rotation:
(1) infraspinatus;
(2) teres minor;
(3) posterior deltoid

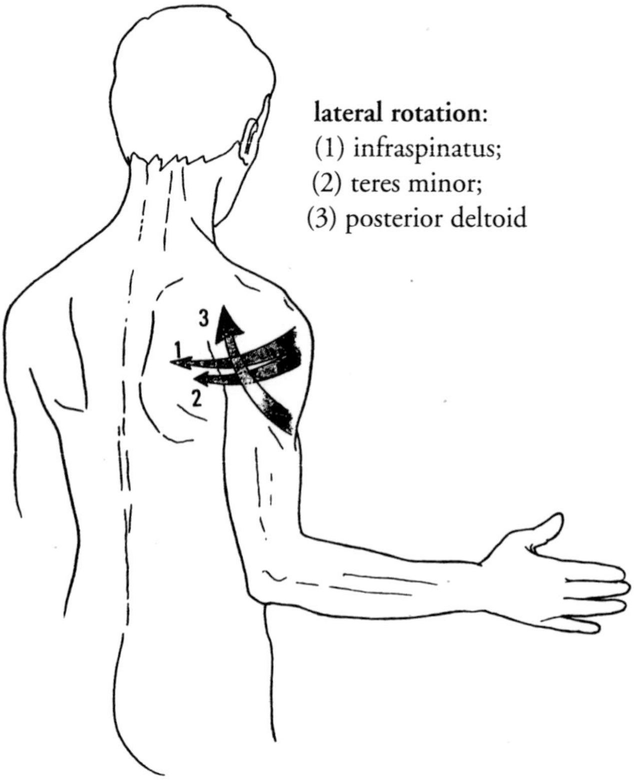

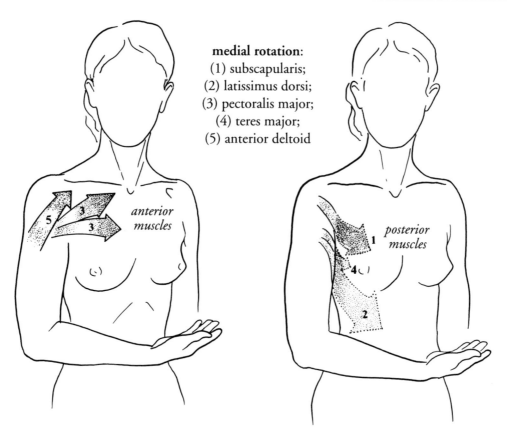

medial rotation:
(1) subscapularis;
(2) latissimus dorsi;
(3) pectoralis major;
(4) teres major;
(5) anterior deltoid

anterior muscles

posterior muscles

In assigning muscle functions as above, we are assuming that the arm starts out in anatomical position. In different positions, functions change and may even be reversed. For example, pectoralis major is a flexor of the arm up to 60°. Beyond 90°, it can no longer move the arm forward or upward; in fact, it begins to function as an extensor, bringing the arm back toward anatomical position. The same idea applies to latissimus dorsi in extension.

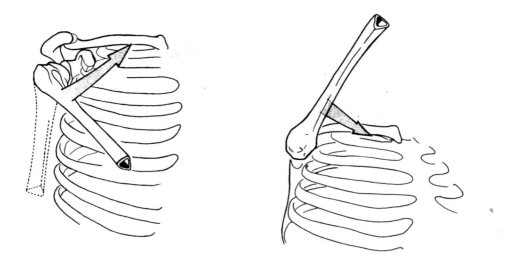

In cases like these, deltoid tends to "take over" at the extremes of movement as other muscles become ineffective.

CHAPTER FOUR

The Elbow

· ·

The forearm contains two bones, the radius and ulna. The ulna is primarily involved in the elbow joint (where it articulates with the humerus), whereas the radius is primarily involved in the wrist joint (where it articulates with the carpal bones). The elbow itself is a simple hinge joint and allows only flexion and extension. In addition, the radius is able to perform the movements called pronation and supination, in which it crosses over the ulna (so that the palm of the hand faces posteriorly) and returns to anatomical position. Although pronation/supination are not, strictly speaking, movements of the elbow joint, we will discuss them in this chapter because the muscles involved are closely related (in some cases identical) to those involved in flexion/extension.

Landmarks

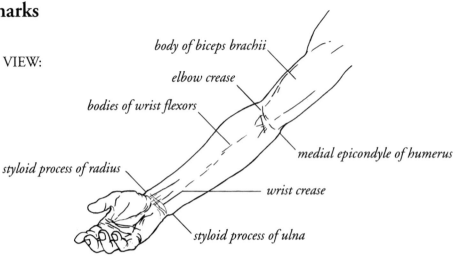

FRONT VIEW:

body of biceps brachii

elbow crease

bodies of wrist flexors

medial epicondyle of humerus

styloid process of radius

wrist crease

styloid process of ulna

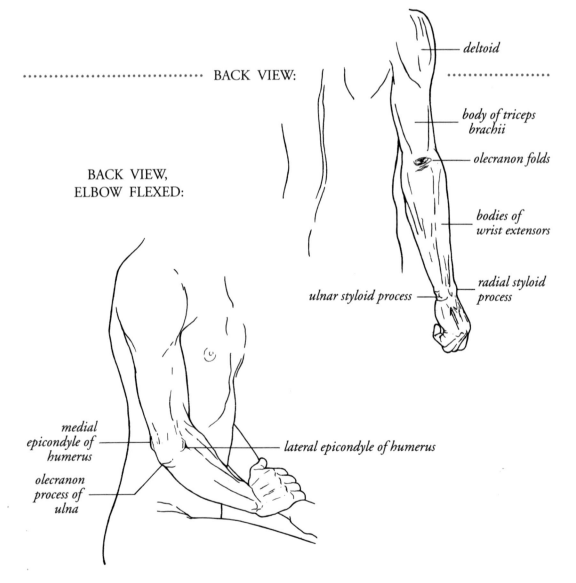

BACK VIEW:

deltoid

body of triceps brachii

olecranon folds

bodies of wrist extensors

BACK VIEW,
ELBOW FLEXED:

ulnar styloid process

radial styloid process

medial epicondyle of humerus

lateral epicondyle of humerus

olecranon process of ulna

Radius and ulna

In anatomical position, the radius is lateral and the ulna is medial. The proximal end of the **radius** is expanded into a **head** which fits against the capitulum of the humerus and the radial notch of the ulna. The **radial tuberosity** is the insertion site for biceps brachii. The shaft is rounded laterally but sharper medially, where it gives rise to an **interosseous membrane** which links it to the shaft of the ulna. Distally, there is a lateral **styloid process** which is easily palpable externally. The **ulnar notch** fits against the distal end of the ulna. The large **articular facet** contributes to the wrist joint by fitting against the scaphoid and lunate (two of the carpal bones).

RADIUS

ULNA

head

trochlear notch

olecranon process

coronoid process

neck

radial notch

radial tuberosity

shaft

shaft

styloid process

ulnar notch

articular facet

"head"

styloid process

At the proximal end of the **ulna** is a large **trochlear notch** which wraps around the trochlea of the humerus. This notch is delimited posteriorly by the **olecranon process** and anteriorly by the **coronoid process**, each of which fits into a corresponding fossa on the humerus. The distal end of the ulna is called the "head" (which is confusing because in all other arm and leg bones the "head" is the proximal end). There is a medial **styloid process**. The distal ulna does not participate in the wrist joint; it is separated from the carpal bones by a fibrocartilage disc.

The distal end of the humerus has a rounded surface (the **capitulum**) laterally, and a pulley-shaped surface (the **trochlea**) medially. Above these, anteriorly, are the **radial fossa** and **coronoid fossa**. Posteriorly, there is an olecranon fossa (not shown). During movement of the elbow joint, the head of the radius (which is slightly concave) slides on the capitulum and the trochlear notch of the ulna slides on the trochlea (below right). In extreme flexion, the radial head and ulnar coronoid process fit into the radial and coronoid fossae of the humerus. In extension, the olecranon process fits into the olecranon fossa, and prevents the elbow from extending past 180°.

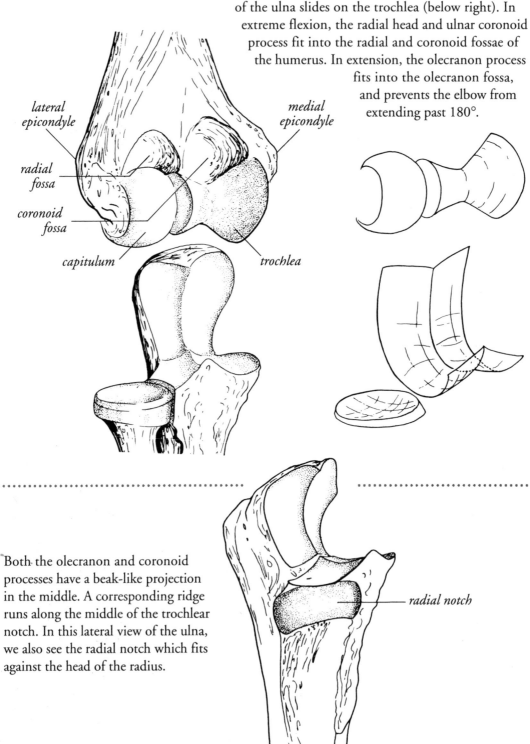

*lateral
epicondyle*

*medial
epicondyle*

*radial
fossa*

*coronoid
fossa*

capitulum

trochlea

Both the olecranon and coronoid processes have a beak-like projection in the middle. A corresponding ridge runs along the middle of the trochlear notch. In this lateral view of the ulna, we also see the radial notch which fits against the head of the radius.

radial notch

Joint capsule

The capsule of the elbow joint, on the anterior humerus, attaches above the radial and coronoid fossae, passing from the lateral to the medial epicondyle. Posteriorly, it attaches above the olecranon fossa. Inferiorly, it attaches on the neck of the radius, the coronoid process, medial surface of the ulna, and the olecranon.

The capsule is taut in front (especially laterally), but loose in back (facilitating flexion).

The proximal radioulnar joint (i.e., between the head of the radius and radial notch of the ulna) should be considered in association with the elbow joint, since the synovial cavities and ligaments of these joints are shared.

The **annular ligament** is a U-shaped ligament wrapping around the head of the radius and attaching on the front and back of the ulna.

— *trochlear notch of ulna*

— *annular ligament*

The **radial collateral ligament** arises from the lateral epicondyle, with two slips inserting on the annular ligament, and the third on the lateral olecranon (left). The **ulnar collateral ligament** arises from the medial epicondyle and attaches below to the medial coronoid process and olecranon (right). Both collateral ligaments allow flexion/extension but prevent any lateral movement of the elbow joint.

Flexion/extension

Our general definition of flexion (in Chapter 1) was "a movement in a sagittal plane which takes a part of the body forward from anatomical position." In the case of the elbow joint, it is more precise to define **flexion** as a movement that decreases the angle between the anterior surfaces of the arm and forearm.

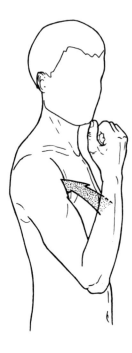

In active flexion, the movement is limited by contact between the bodies of the muscles involved. In passive flexion, the angle may become slightly smaller since the muscles are more compressible.

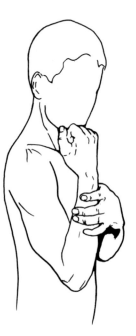

Extension of the elbow is a return from flexion to anatomical position, i.e., increase in the angle between the arm and forearm.

Generally, increase of the angle past 180° is impossible due to contact between the olecranon process and its fossa. In a few exceptional individuals, the angle can become slightly larger than 180°; this is called "recurvation."

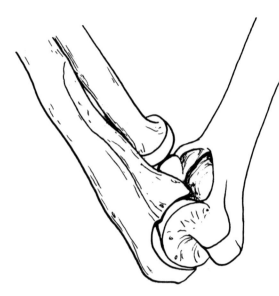

In extreme flexion, the radial head and coronoid process fit against their corresponding fossae on the distal humerus.

In extension, the olecranon process fits against its fossa.

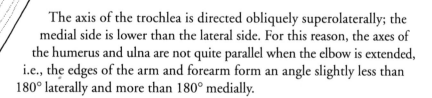

The axis of the trochlea is directed obliquely superolaterally; the medial side is lower than the lateral side. For this reason, the axes of the humerus and ulna are not quite parallel when the elbow is extended, i.e., the edges of the arm and forearm form an angle slightly less than 180° laterally and more than 180° medially.

Muscles of flexion/extension

Brachialis arises from the anterior surface of the distal humerus and inserts on the coronoid process.

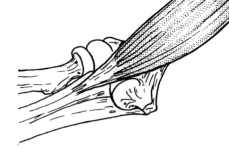

It is a major flexor of the elbow.

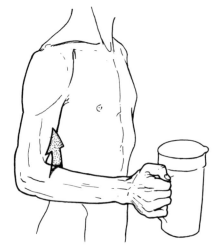

Brachioradialis originates from a lateral ridge on the distal humerus, and inserts via a long tendon on the base of the radial styloid process.

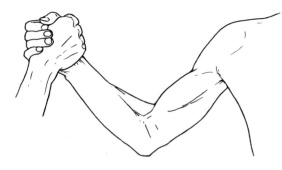

It flexes the elbow, and is particularly effective if the radius is partially pronated.

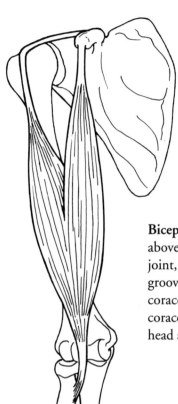

Biceps brachii has two origins. The long head arises from a tubercle above the glenoid cavity of the scapula, travels through the shoulder joint, between the greater and lesser tubercles, and along the bicipital groove before merging with the body. The short head arises from the coracoid process of the scapula (where it is fused with the origin of coracobrachialis), and the muscular fibers meet those from the long head about halfway down the humerus.

Biceps inserts on the radial tuberosity and bicipital aponeurosis (see p. 123). This muscle and brachialis are the primary elbow flexors. However, because it inserts on the posterior aspect of the radial tuberosity, biceps is also an important supinator of the radius.

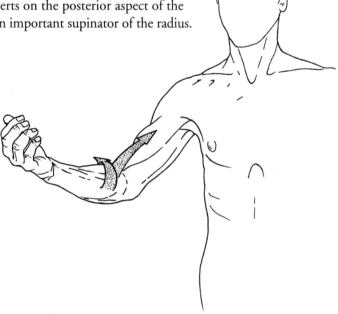

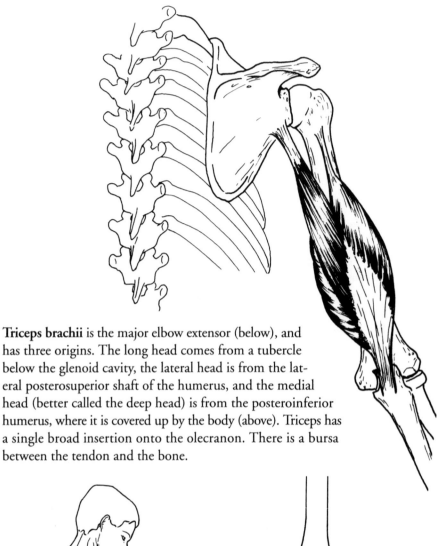

Triceps brachii is the major elbow extensor (below), and has three origins. The long head comes from a tubercle below the glenoid cavity, the lateral head is from the lateral posterosuperior shaft of the humerus, and the medial head (better called the deep head) is from the posteroinferior humerus, where it is covered up by the body (above). Triceps has a single broad insertion onto the olecranon. There is a bursa between the tendon and the bone.

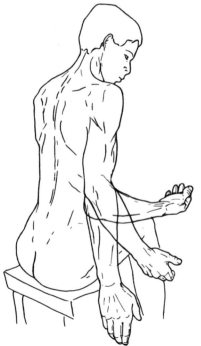

Anconeus is a small muscle running from the lateral epicondyle to the lateral olecranon and superior ulna. It assists the triceps in extension, and also plays a small role in pronation (see p. 144).

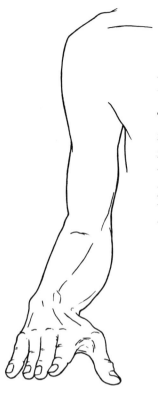

Pronation/supination

These movements involve changes in the relationship between the radius and ulna, not the ulna and humerus. In pronation, the radius crosses over the ulna so that the palm faces posteriorly (left). In supination, the radius and ulna become parallel, and the palm faces anteriorly (right).

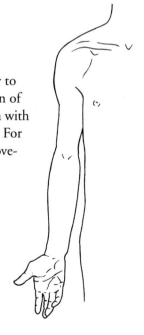

If the elbow is fully extended, it is easy to confuse pronation with medial rotation of the shoulder joint (left), and supination with lateral rotation of the shoulder (right). For this reason, it is best to study these movements with the elbow flexed.

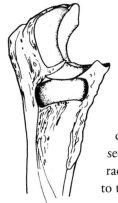

What happens at the proximal radioulnar joint during pronation? As we have seen, the ulna has a concave radial notch just inferolateral to the coronoid process.

The annular ligament attaches at the front and back of this notch.

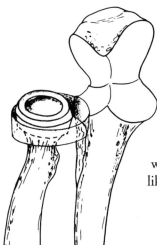

The notch and ligament are both lined with synovial membrane, and form a ring-like structure around the radial head.

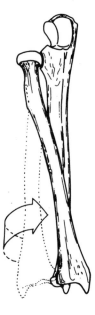

This ring allows easy rotation of the radial head during pronation.

The ring is somewhat funnel-shaped (wider at the top than the bottom), and helps resist downward movement of the radius when the forearm is supporting a heavy weight.

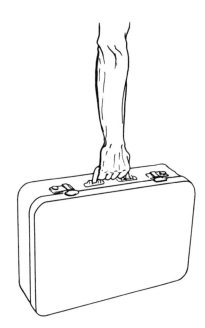

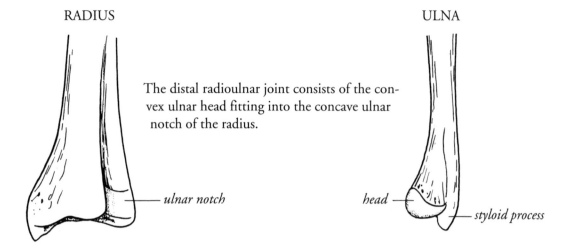

RADIUS

ULNA

The distal radioulnar joint consists of the convex ulnar head fitting into the concave ulnar notch of the radius.

ulnar notch

head

styloid process

RADIUS   ULNA

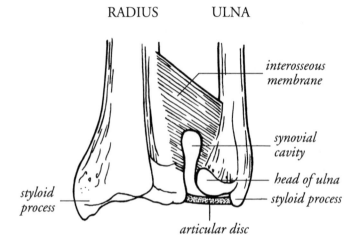

interosseous membrane

synovial cavity

head of ulna

styloid process

styloid process

articular disc

These articular surfaces are covered with hyaline cartilage and separated by a thin layer of synovial fluid. This synovial cavity is not continuous with the radiocarpal joint cavity. There is a fibrocartilaginous **articular disc** connecting the base of the ulnar styloid process to the edges of the ulnar notch (left).

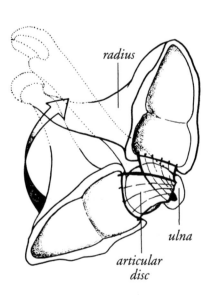

radius

ulna

articular disc

This disc provides the strongest attachment between the distal radius and ulna. It "sweeps" the ulnar head during pronation and supination (left). During pronation, the posterior portion of the disc becomes taut; in supination, the anterior portion becomes taut. The **interosseous membrane** which connects the shafts of the radius and ulna has diagonal fibers oriented in both directions (right). This membrane is relaxed in pronation and taut in supination, and thus acts as a brake of supination. It also prevents longitudinal displacement of the two bones.

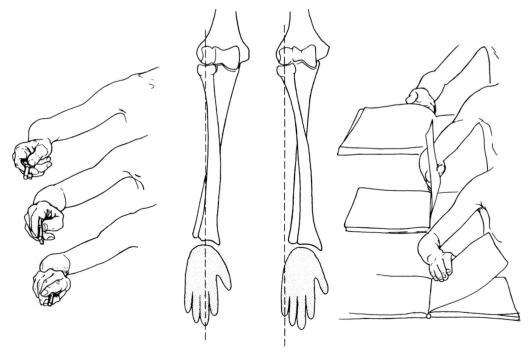

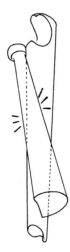

There are two slightly different types of pronation. In the first (e.g., turning a key), the axis for movement of the hand passes through the middle finger, and the ulna moves slightly in conjunction with the radius. Anconeus is involved in this movement. In the second (e.g., flipping the page of a book), the axis of the hand passes through the fifth finger, and the ulna remains fixed.

The ulna and radius, in anatomical position, are both concave anteriorly.

This curvature allows the radius to cross over the ulna during pronation.

If both bones were straight, they would contact each other too soon and normal pronation would be impossible.

Fractures or other injuries can alter these curvatures and thereby interfere with pronation. This is a point of concern in certain disciplines (e.g., martial arts) involving unusual stresses on the forearm.

Muscles of pronation/supination

Pronator teres arises from the medial epicondyle of the humerus and coronoid process of the ulna, and inserts on the midlateral surface of the radius. It is the major pronator of the forearm, and assists in flexion of the elbow.

Pronator quadratus is a square-shaped muscle running between the anterior surfaces of the distal ulna and radius. By contracting, it pulls the radius across the ulna in pronation.

pronator teres —

pronator
quadratus

Brachioradialis was described on page 138. Although primarily an elbow flexor, it can assist in the initial stage of pronation from a supinated position. Conversely, it can also assist in the initial stage of supination from a pronated position! In other words, it tends to move the radius to a position intermediate between complete supination and complete pronation.

Biceps brachii was described on page 139. Besides its role as an elbow flexor, it is also an important forearm supinator because of the location of its insertion (on the posterior aspect of the radial tuberosity); i.e., it "uncrosses" the upper radius from a pronated position.

Supinator originates in two layers. The superficial layer is from the lateral epicondyle of the humerus; the deep layer is from the "supinator ridge" located just below the posterior radial notch of the ulna. This muscle wraps around the radius, inserting between the neck and the insertion of pronator teres (right). As you may have guessed by its name, it supinates the forearm; it cooperates with biceps brachii in returning the proximal radius from a pronated position to anatomical position.

supinators

pronator

The radius can be visualized as having a "supinating curve" (with biceps brachii and supinator inserting near the top), and a "pronating curve" (with pronator teres inserting near the top). By contracting alternately, these muscles turn the radius like a crank.

CHAPTER FIVE

The Wrist & Hand

....

The radius articulates with the carpal bones at the wrist joint. The carpals join distally to the metacarpal bones, which in turn join to the phalanges (finger bones). The hand consists of the metacarpals and phalanges together with the surrounding muscles, vessels, nerves, skin, etc.

The hand is a versatile "tool" capable of tasks ranging from fine precision (playing piano, threading a needle) to great strength (swinging a sledgehammer, unscrewing a stubborn jar lid). The success of humans as a species is due in large part to our ability to grasp and manipulate objects. The strong, opposable thumb is crucial in this respect, and is one adaptation which allowed humans to take the evolutionary "quantum leap" beyond their primate ancestors.

Landmarks

ANTERIOR (PALMAR) VIEW:

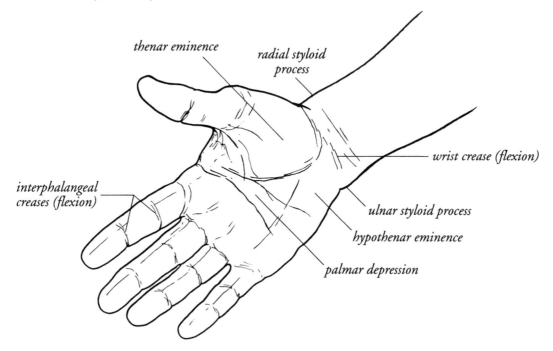

thenar eminence

radial styloid process

wrist crease (flexion)

interphalangeal creases (flexion)

ulnar styloid process

hypothenar eminence

palmar depression

POSTERIOR (DORSAL) VIEW: ....

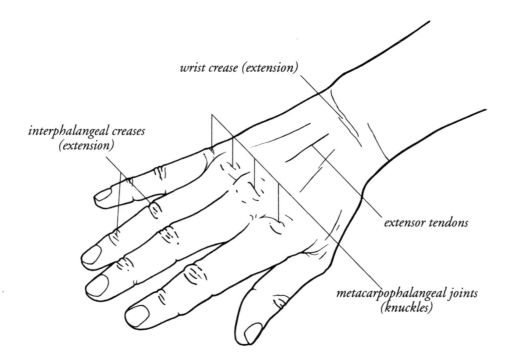

wrist crease (extension)

interphalangeal creases (extension)

extensor tendons

metacarpophalangeal joints (knuckles)

Bones

Bones of the wrist and hand can be divided into three groups.

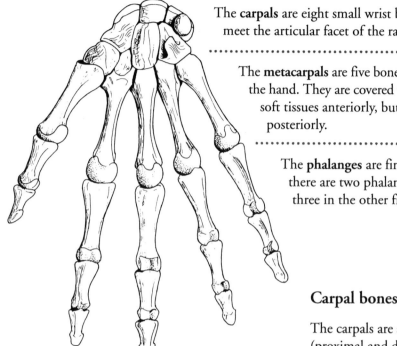

The **carpals** are eight small wrist bones; two of them meet the articular facet of the radius.

· ·

The **metacarpals** are five bones located in the palm of the hand. They are covered with muscles and other soft tissues anteriorly, but can be easily palpated posteriorly.

· ·

The **phalanges** are finger bones. Note that there are two phalanges in the thumb, and three in the other fingers.

Carpal bones

The carpals are arranged in two rows (proximal and distal), each containing four bones.

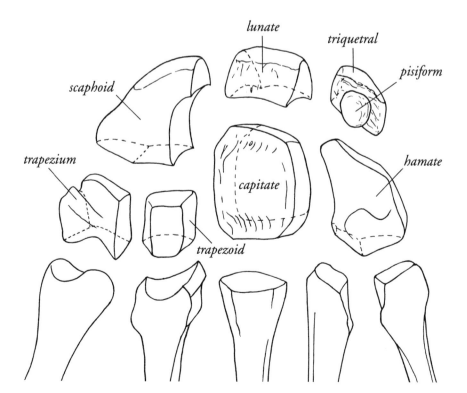

- The **scaphoid** articulates superiorly with the radius and inferiorly with the trapezium and trapezoid. In most wrist fractures, only the scaphoid is involved.

- The **lunate** articulates superiorly with the radius and inferiorly with the capitate.

- The **triquetral** articulates superiorly with the articular disc which runs between the ulnar styloid process and the radius (see page 143). Inferiorly, it contacts the hamate.

- The **pisiform** is a small round bone which sits on the anterior surface of the triquetral and can be palpated externally. It does not articulate with the forearm nor with the hamate, but does serve for attachment of some ligaments.

- The **trapezium** has a sharp anterior crest. It joins metacarpal I (the metacarpals and fingers are designated by roman numerals I to V, from lateral to medial).

- The **trapezoid** is the most symmetrical of the carpal bones, being shaped like a pyramid with the top cut off. It articulates with metacarpal II.

- The **capitate** is the largest carpal and has an anterior tubercle. It articulates primarily with metacarpal III and has two facets on the inferior corners which contact metacarpals II and IV.

- The **hamate** has a prominent anterior projection called the "hook" which can sometimes be palpated in the hypothenar eminence distal to the pisiform. The inferior surface of the hamate has two facets oriented in different directions which articulate with metacarpals IV and V.

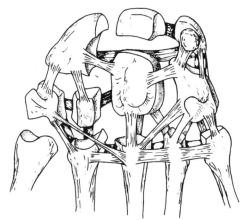

Where they contact each other, the carpals are covered by hyaline cartilage and form joints of the "gliding" type (see page 9). There are numerous small ligaments binding the carpals to one another and to the metacarpals.

Together, the carpals form an anteriorly-concave arch. The sides of the arch consist laterally of the scaphoid and trapezium, and medially of the triquetral, pisiform, and hamate.

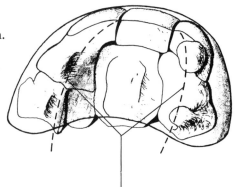

anteriorly-concave arch

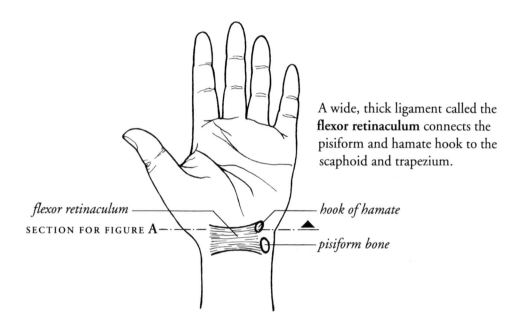

A wide, thick ligament called the **flexor retinaculum** connects the pisiform and hamate hook to the scaphoid and trapezium.

flexor retinaculum

SECTION FOR FIGURE **A**

hook of hamate

pisiform bone

FIGURE A: ••

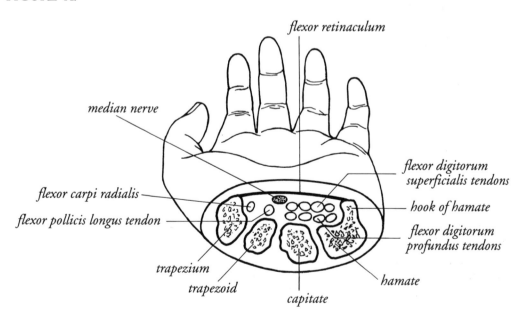

flexor retinaculum

median nerve

flexor carpi radialis

flexor pollicis longus tendon

flexor digitorum superficialis tendons

hook of hamate

flexor digitorum profundus tendons

trapezium

trapezoid

capitate

hamate

The carpal arch and the flexor retinaculum delimit a narrow space called the **carpal tunnel** through which the median nerve and tendons from many hand flexor muscles pass.

Posteriorly, there is a corresponding **extensor retinaculum** to keep the tendons of extensor muscles in place. The extensor retinaculum attaches laterally to the distal radius, and medially to the ulnar styloid, triquetral, and pisiform.

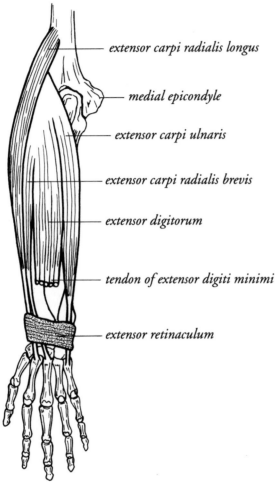

extensor carpi radialis longus

medial epicondyle

extensor carpi ulnaris

extensor carpi radialis brevis

extensor digitorum

tendon of extensor digiti minimi

extensor retinaculum

Superiorly, the scaphoid, lunate, and tri-quetral present a large convex surface which articulates with the smaller concave surface presented by the distal radius and articular disc to form an ellipsoid joint.

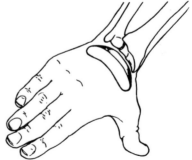

Metacarpals and phalanges

There are a total of five metacarpals and 14 phalanges (the thumb has two phalanges, the other fingers have three each).

Each of these bones has a base (proximal), shaft, and head (distal).

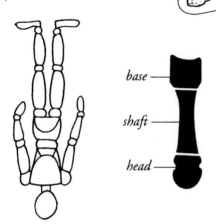

The base of each **metacarpal** is roughly quadrangular, with facets for articulation with a carpal and the adjacent metacarpals.

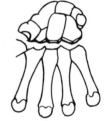

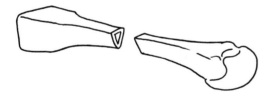

The shaft is roughly triangular, as in many long bones. The head is convexly rounded and has a small tubercle on each side for attachment of tendons.

The proximal **phalanx** of each finger has a concavely rounded base for articulation with the metacarpal, and a pulley-shaped head. The base of the middle phalanx is concave but with a median crest to match the shape of the head of the proximal phalanx. The head of the distal phalanx is flared to support the fingernail and fleshy anterior part of the fingertip.

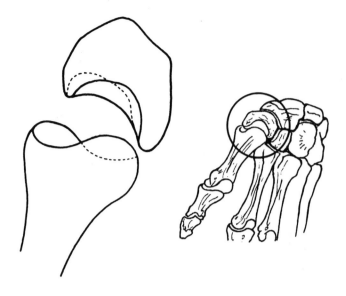

Where the trapezium meets metacarpal I, both articular surfaces are saddle-shaped, which allows good range of movement in many directions, like a cowboy sitting on a horse's saddle.

Thus, the thumb can be moved to oppose the other fingers, allowing grasping and manipulation of objects.

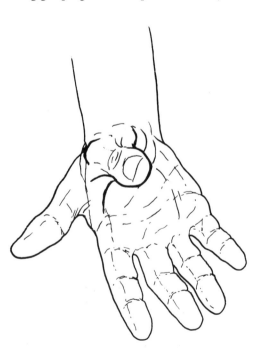

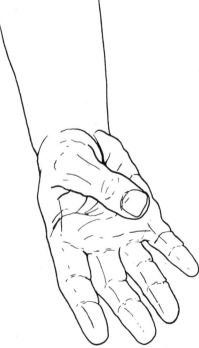

Joints

Carpal joints

The ellipsoid joint between distal radius, articular disc, and proximal carpals is called the **radio-carpal joint**. The movements of the articular disc during pronation/supination were described on page 143. There are two **interosseous ligaments** connecting the lunate to the scaphoid and triquetral; these help close off the radiocarpal joint cavity. The **medial collateral ligament** runs from the ulnar styloid to the pisiform and triquetral, and the **lateral collateral ligament** from the radial styloid to the scaphoid.

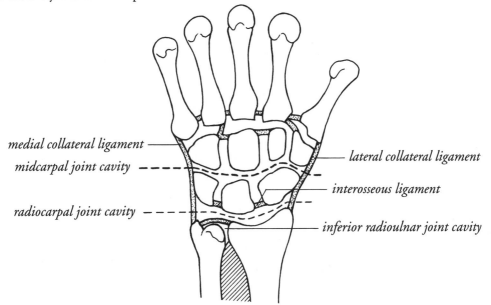

medial collateral ligament

midcarpal joint cavity

radiocarpal joint cavity

lateral collateral ligament

interosseous ligament

inferior radioulnar joint cavity

Various smaller ligaments reinforce the joint anteriorly and posteriorly (below).

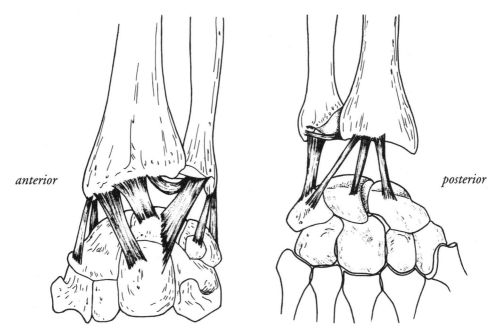

anterior

posterior

The joint capsule is generally slack in the superoinferior direction, but taut from side to side. It is important to note that the radiocarpal joint allows only flexion/extension and abduction/adduction. What may appear to be rotation of this joint is actually pronation/supination of the forearm. You can verify this by clasping your left hand around the right forearm, and attempting to "rotate" the right wrist without moving the radius.

The **midcarpal joint** is the joint between the proximal and distal rows of carpals, actually a combination of small gliding "subjoints" between individual bones. The joint capsule here is slightly thicker posteriorly than anteriorly, and the enclosed synovial membrane is continuous. There are interosseous ligaments joining the capitate to the hamate and trapezoid (see page 157).

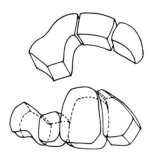

Due to the irregular shape of the bones (particularly the scaphoid), this joint includes several bends.

Carpometacarpal joints · · · · · · · · · · · · · ·

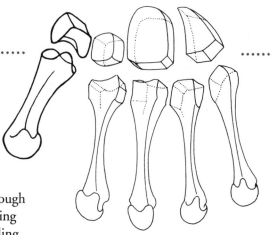

These are the joints between the distal row of carpals and the metacarpals (right).

Carpometacarpal joint I (thumb joint) is a well-developed saddle joint (see page 10), and has good range of movement. The base of metacarpal I is rotated 90° from that of metacarpal II. Joints II through V are slightly saddle-shaped but closer to being gliding joints. They allow slight sliding/gliding and flexion/extension movements (below).

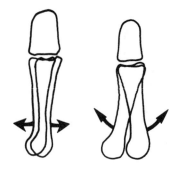

The range of these movements increases progressively from metacarpal II through V. As a result of the anterior curvature of the carpals, the plane of carpometacarpal joints IV and V is oblique to that of joints II and III.

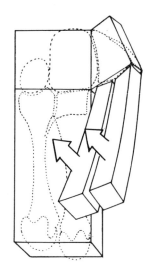

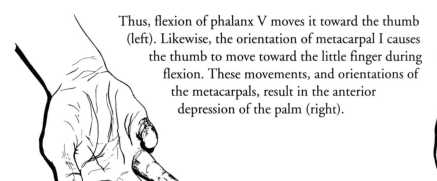

Thus, flexion of phalanx V moves it toward the thumb (left). Likewise, the orientation of metacarpal I causes the thumb to move toward the little finger during flexion. These movements, and orientations of the metacarpals, result in the anterior depression of the palm (right).

Some important connective tissue structures should be mentioned here. The **palmar aponeurosis**, formed from the deep fascia of the anterior hand, extends from the flexor retinaculum to the four fingers. It is attached to the skin of the palm and is thickest centrally. From this aponeurosis arise two extensions which pass posteriorly: the **medial septum** which attaches to metacarpal V, and the **lateral septum** which attaches to metacarpal I. In cross-section (below), we see that these septa divide the palm into three compartments: a medial compartment containing mostly hypothenar muscles, a lateral compartment containing thenar muscles, and a central compartment containing flexor tendons, their sheaths, and considerable loose tissue which keeps the tendons in place.

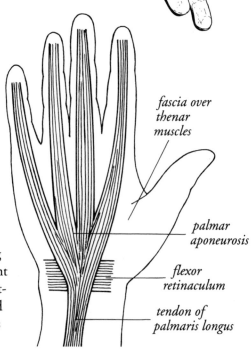

fascia over thenar muscles

palmar aponeurosis

flexor retinaculum

tendon of palmaris longus

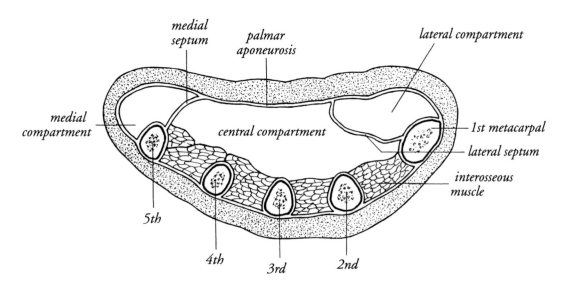

medial septum

palmar aponeurosis

lateral compartment

medial compartment

central compartment

1st metacarpal

lateral septum

interosseous muscle

5th

4th

3rd

2nd

Metacarpophalangeal joints

These are essentially hinge joints allowing good flexion/extension.

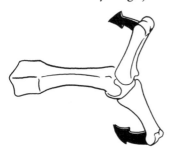

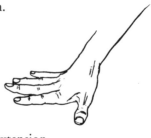

Range of passive extension is greater than that of active extension.

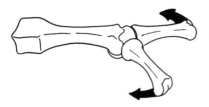

Limited abduction/ adduction and rotation are also possible.

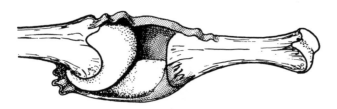

The joint capsule is slack at the front and back, taut at the sides, and reinforced on the palmar surface by the **palmar ligament**, a dense band of fibrocartilaginous tissue. This ligament is loosely attached to the metacarpal head and densely attached to the base of the proximal phalanx.

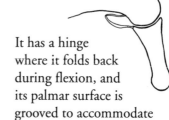

It has a hinge where it folds back during flexion, and its palmar surface is grooved to accommodate the flexor tendons.

There are two **collateral ligaments** on either side of the joint. Since they originate from the dorsal side of the meta- carpal head, which is somewhat narrower than the palmar side, these ligaments are slack in extension (left) and taut in flexion (right). Consequently, movements of abduction/adduction and rotation are impossible when the joint is in full flexion.

When the metacarpophalangeal joints are in extension or slight flexion, passive abduction/ adduction and rotation allow the hand to adapt itself to grasp a variety of shapes (left). When these joints are in a more flexed position, they become less flexible but also more stable, which is helpful for feats requiring strength or force (right).

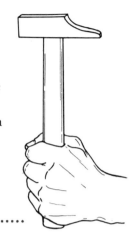

Interphalangeal joints

These are smaller-scale versions of the metacarpophalangeal joints, and have similar capsules and collateral ligaments. Movements other than flexion/ extension are more restricted due to the pulley-like shape of the articular surfaces.

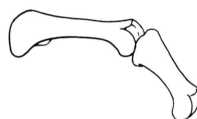

Between the proximal and middle phalanx, extension is usually limited to 180° (below).

Between the middle and distal phalanx, extension past 180° is possible to a varying extent, depending on the individual (below).

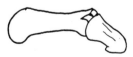 

Thumb joints

Metacarpophalangeal joint I differs from II through V in a few respects:

- it is more massive;
- the capsule is not as taut laterally, and allows some axial rotation;
- two small sesamoid bones are embedded in the palmar fascia, and serve for tendon attachment.

The interphalangeal joint is similar to those of fingers II through V except for being more massive (right).

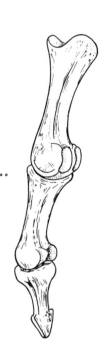

Movements

Movements of the wrist

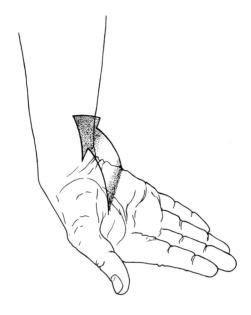

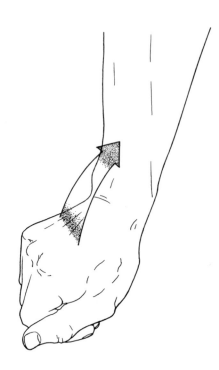

In **flexion** of the wrist, the palm moves closer to the anterior surface of the forearm.

In **extension** of the wrist, the posterior surfaces of the hand and forearm move closer together.

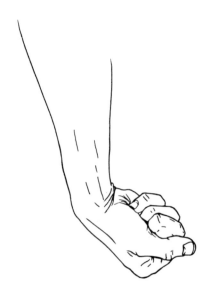

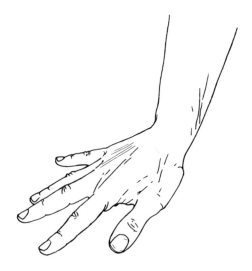

The fingers tend to elongate during this movement, due to tightening of the extensor tendons. You can feel this tightening on the back of the hand when flexing the fingers.

In this case, the fingers tend to flex, due to tightening of the flexor tendons. You can feel these tendons on the palm when extending the fingers.

Flexion and extension of the wrist have roughly the same range of motion, usually 80° to 90° from anatomical position.

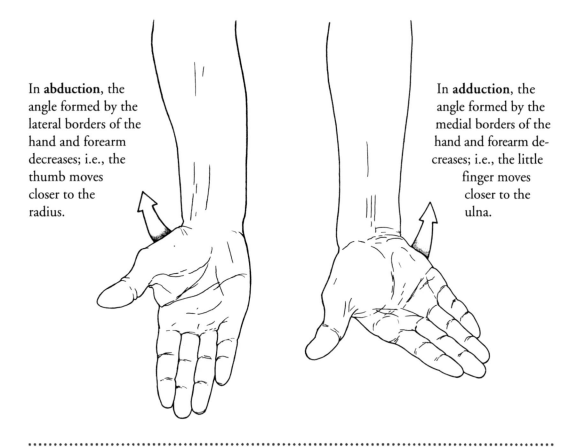

In **abduction**, the angle formed by the lateral borders of the hand and forearm decreases; i.e., the thumb moves closer to the radius.

In **adduction**, the angle formed by the medial borders of the hand and forearm decreases; i.e., the little finger moves closer to the ulna.

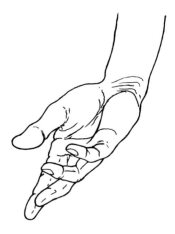

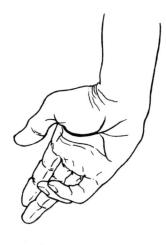

The range of motion for adduction is greater than that of abduction. Because of the muscles involved, adduction tends to be combined with flexion...

...whereas abduction tends to occur together with extension.

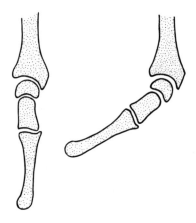

Wrist flexion is primarily a function of the radiocarpal joint.

Extension involves the midcarpal joint to a greater extent, since the posterior border of the distal radius limits extension at the radiocarpal joint.

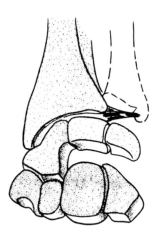

In abduction, the scaphoid moves closer to the radius. This movement is limited by the radial styloid. Medially, the proximal and distal rows of carpals move apart.

In adduction, the triquetral moves closer to the ulna, while the scaphoid moves away from the radius. Movement is less restricted here because the ulnar styloid is not as prominent as the radial styloid.

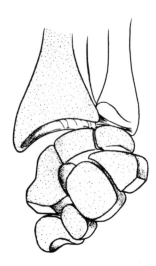

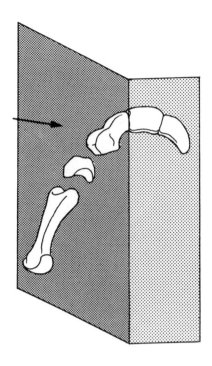

Movements of the thumb

Due to the progressively oblique orientation of the scaphoid, trapezium, and metacarpal I, the base of this metacarpal is rotated nearly 90° from that of metacarpal II (left). Thus, in a hand at rest, the thumb faces the other fingers at a right angle (below).

Movements of the thumb (i.e., metacarpal I) must be defined differently from those of the other fingers.

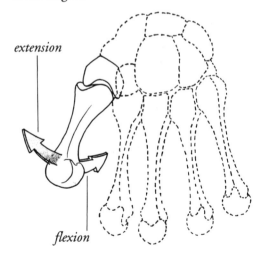

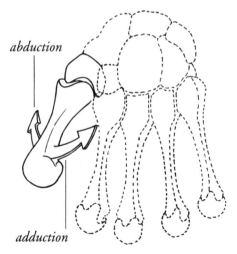

In **extension**, the metacarpal moves posterolaterally, while in **flexion** it moves anteromedially, closer to the palm.

In **abduction**, it moves anterolaterally, while in **adduction** it moves posteromedially.

· ·

The capsule of carpometacarpal joint I is slack, allowing some axial rotation in addition to the movements described above and further enhancing the thumb's mobility.

Movements of the carpometacarpal, metacarpophalangeal, and interphalangeal joints for fingers II through V were described in the preceding section.

Muscles

The major muscles responsible for movements of the wrist and fingers have their bodies around the radius and ulna, and long tendons leading to the insertion bone(s). However, some small intrinsic hand muscles have their bodies around the metacarpals or phalanges.

We shall describe the muscles on the basis of functional groups. Usually, muscles with similar function also have similar location.

Muscles moving the wrist ···

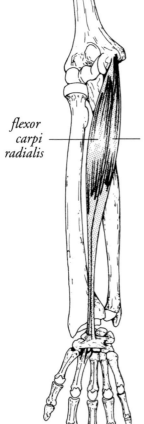

flexor carpi radialis

palmaris longus

Flexor carpi radialis arises from the "common flexor origin" at the medial epicondyle of the humerus. Its tendon passes along the groove of the trapezium and inserts on the bases of metacarpals II and III (left). It flexes and abducts the wrist, acting on both the radiocarpal and midcarpal joints (below).

Palmaris longus arises from the common flexor origin and inserts on the flexor retinaculum and palmar aponeurosis (left). It is a weak wrist flexor (below) and takes no part in abduction or adduction because of its central location. It is absent in some individuals.

Flexor carpi ulnaris runs from the common flexor origin along the medial ulna, and inserts on the pisiform, which transmits its force to the hamate hook and base of metacarpal V via ligaments (left). It flexes and adducts the wrist (right).

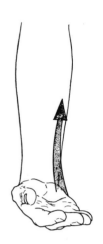

Extensor carpi radialis longus originates from the lateral epicondyle ("common extensor origin") and supracondylar ridge of the humerus. Its tendon passes under the extensor retinaculum and inserts on the posterior base of metacarpal II (below left). **Extensor carpi radialis brevis** arises from the common extensor origin and inserts on the posterior base of metacarpal III (below right).

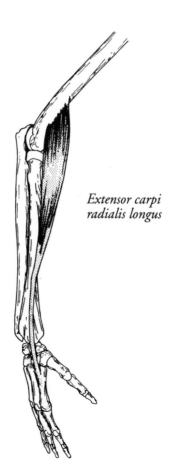

Extensor carpi radialis longus

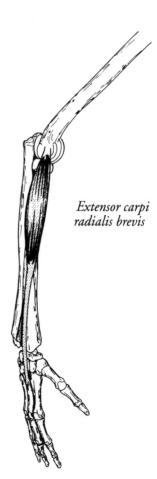

Extensor carpi radialis brevis

Action: both muscles extend and abduct the wrist

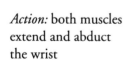

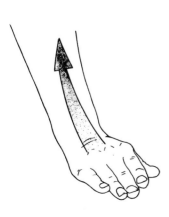

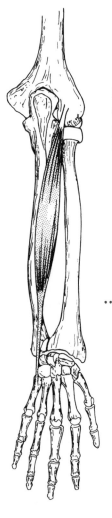

Extensor carpi ulnaris originates from the common extensor origin and the posterior border of the ulna, passes under the extensor retinaculum, and inserts on the posterior base of metacarpal V.

Action: extends and adducts the wrist

Extrinsic muscles moving the hand

Flexor digitorum profundus has a broad origin on the anterior and medial ulna, and medial half of the interosseous membrane which connects the ulna and radius. It splits into four tendons which pass through the carpal tunnel and insert on the distal phalanges of fingers II through V.

Action: flexion of *all* inter-phalangeal joints of fingers II through V, as well as the metacarpophalangeal and (in some situations) wrist joints

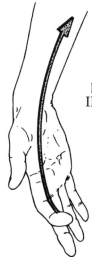

Flexor digitorum superficialis has two heads: one from the common flexor origin, the elbow joint capsule, and the coronoid process of the ulna; the other from the anterior surface of the radius. It splits into four tendons which pass through the carpal tunnel (superficial to the tendons of flexor digitorum profundus, of course), split into "Y" shapes (to accommodate passage of other flexor tendons), and insert bilaterally on the middle phalanges of fingers II through V.

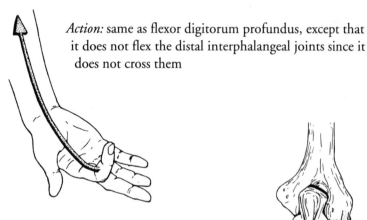

Action: same as flexor digitorum profundus, except that it does not flex the distal interphalangeal joints since it does not cross them

Extensor digitorum arises from the common extensor origin and deep fascia of the forearm. It splits into four tendons which pass under the extensor retinaculum (right). Each tendon in turn splits into three bands, of which the central band inserts on the posterior base of the proximal and middle phalanges, while the two lateral bands reunite at the base of the distal phalanx (below).

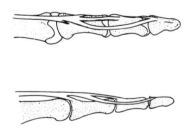

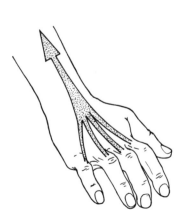

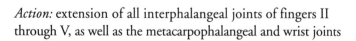

Action: extension of all interphalangeal joints of fingers II through V, as well as the metacarpophalangeal and wrist joints

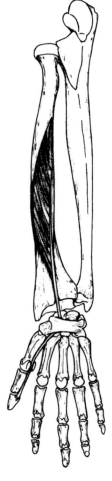

Flexor pollicis longus originates from the anterior radius and lateral part of the interosseous membrane. Its tendon passes through the carpal tunnel and inserts on the distal phalanx of the thumb (left).

Action: flexion of interphalangeal joint I, metacarpophalangeal joint I, carpometacarpal joint I, and the wrist

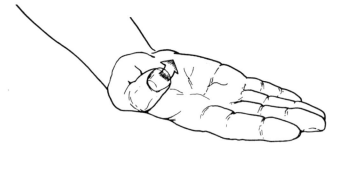

Abductor pollicis longus arises from the posterior surfaces of the ulna, radius, and interosseous ligament, inferior to supinator. The tendon passes under the extensor retinaculum and inserts on the lateral base of metacarpal I (right).

Action: abduction and extension of carpometacarpal joint I

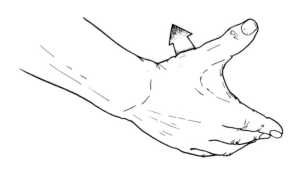

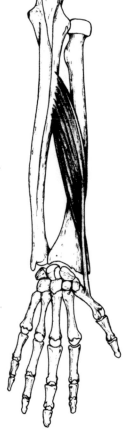

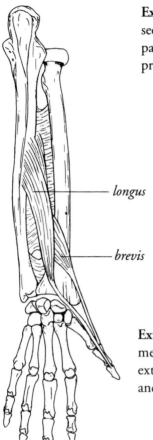

Extensor pollicis brevis originates on the posterior radius and interosseous membrane, inferior to abductor pollicis longus. The tendon passes under the extensor retinaculum and inserts on the base of the proximal phalanx of the thumb (left).

longus

brevis

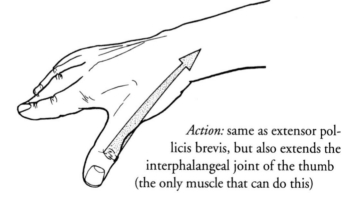

Action: extension of metacarpophalangeal and carpometacarpal joints of the thumb

Extensor pollicis longus arises on the posterior ulna and interosseous membrane, inferior to abductor pollicis longus and superior to extensor indicis. The tendon passes under the extensor retinaculum and inserts on the base of the distal phalanx of the thumb (left).

Action: same as extensor pollicis brevis, but also extends the interphalangeal joint of the thumb (the only muscle that can do this)

When the thumb is fully extended, a depression known as the "anatomical snuffbox" can be seen at the posterior base of the thumb. It is bordered laterally by the tendons of abductor pollicis longus and extensor pollicis brevis, and medially by the tendon of extensor pollicis longus.

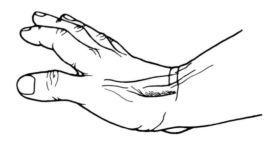

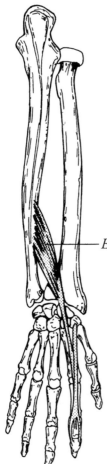

Extensor indicis arises from the posterior ulna and interosseous membrane, below the origin of extensor pollicis longus (left). Its tendon joins that of extensor digitorum leading to finger II (index finger), and assists this muscle in extension of this finger.

— *Extensor indicis*

Extensor digiti minimi originates from the common extensor origin (right), and its tendon joins that of extensor digitorum leading to finger V (little finger), where it assists in extension.

tendon —

Intrinsic muscles moving the hand ·······················

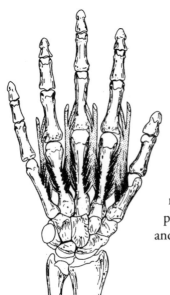

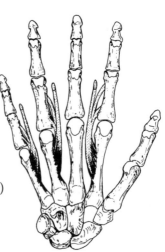

The **interossei** are small muscles originating from the metacarpals and inserting on the phalanges. There are four dorsal (left) and three palmar (right) interossei.

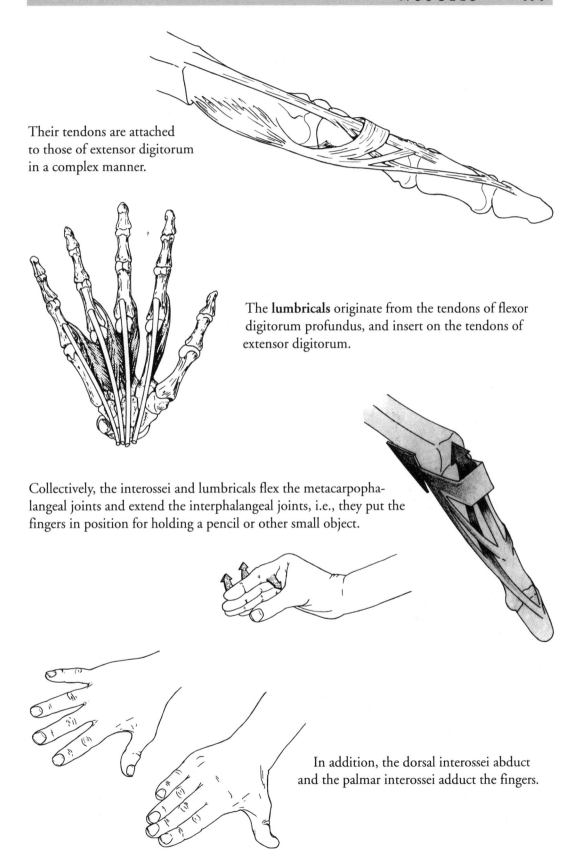

Their tendons are attached to those of extensor digitorum in a complex manner.

The **lumbricals** originate from the tendons of flexor digitorum profundus, and insert on the tendons of extensor digitorum.

Collectively, the interossei and lumbricals flex the metacarpophalangeal joints and extend the interphalangeal joints, i.e., they put the fingers in position for holding a pencil or other small object.

In addition, the dorsal interossei abduct and the palmar interossei adduct the fingers.

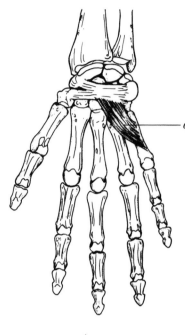

The bodies of the **hypothenar muscles** provide the bulk of the hypothenar eminence on the medial side of the palm. **Opponens digiti minimi** originates from the flexor retinaculum and hamate hook, and inserts on the medial surface of metacarpal V.

opponens digiti minimi

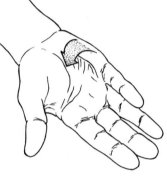

Action: helps move the little finger toward the thumb (for grasping) and create the curvature of the palm

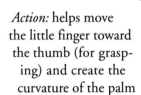

abductor digiti minimi
flexor digiti minimi

Flexor digiti minimi has the same origin as opponens, and inserts on the base of the proximal phalanx of finger V. **Abductor digiti minimi** arises from the pisiform and flexor retinaculum, and inserts in the same place as flexor digiti minimi.

The actions of these two muscles are obvious from their names.

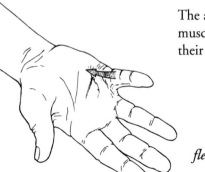

flexion

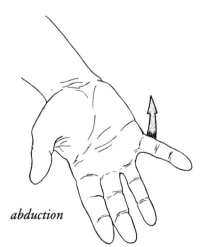

abduction

The **thenar muscles** move the thumb and its metacarpal, and their bodies make up the prominent thenar eminence at the base of the thumb. **Adductor pollicis** lies deep to the flexor tendons in the palm and has two origins: one from the shaft of metacarpal III, the other from the capitate bone and adjacent ligaments. It inserts on the medial base of the proximal phalanx of the thumb, and the medial sesamoid bone located at metacarpophalangeal joint I.

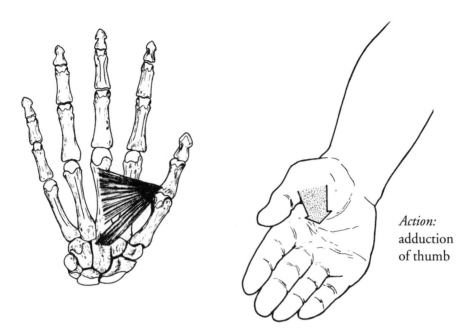

Action: adduction of thumb

Flexor pollicis brevis lies medial to abductor pollicis brevis (see below) and arises from the flexor retinaculum and trapezium. It inserts on the lateral base of the proximal phalanx of the thumb, and the lateral sesamoid bone located at metacarpophalangeal joint I.

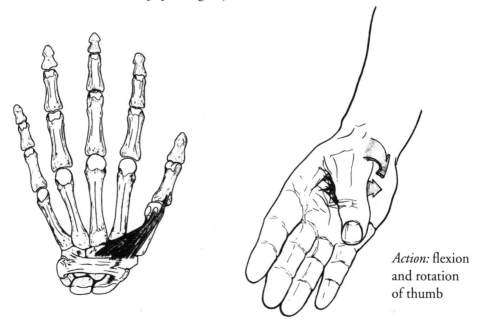

Action: flexion and rotation of thumb

Opponens pollicis has an origin similar to flexor pollicis brevis and lies deep to this muscle. It inserts on the lateral shaft of metacarpal I.

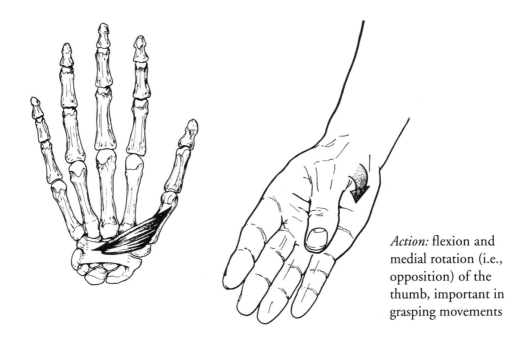

Action: flexion and medial rotation (i.e., opposition) of the thumb, important in grasping movements

Abductor pollicis brevis arises from the flexor retinaculum, scaphoid, and trapezium, and inserts on the lateral base of the proximal phalanx of the thumb next to flexor pollicis brevis.

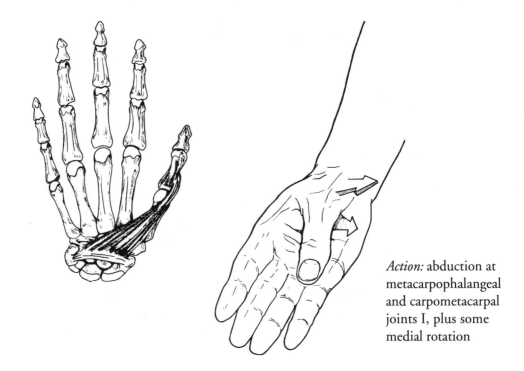

Action: abduction at metacarpophalangeal and carpometacarpal joints I, plus some medial rotation

CHAPTER SIX

The Hip & Knee

The hip is the ball-and-socket joint between the pelvis and femur (upper leg bone). It is surrounded by thick muscles, and therefore difficult to palpate or localize. The stability and powerful musculature of this joint are essential for standing, walking, running, etc.

Dance and most other physical disciplines require good range of motion (ROM) at the hip. Restrictions of ROM here are common, and typically affect nearby structures such as the lower back and knee joint. To work with the hip, especially when there are problems, we need detailed knowledge of its structure and function.

The knee, which connects the distal femur to the tibia (larger of the two lower leg bones), is primarily a hinge joint capable of flexion and extension. Its stability is not due to bone structure (the articulating surfaces here, in contrast to those of the elbow, do not fit snugly), but rather to the arrangement of ligaments and muscles. The knee receives considerable stress, both from above (body weight, gravity) and below (impact from running, etc.) Sports-related injuries to the knee are common.

Landmarks

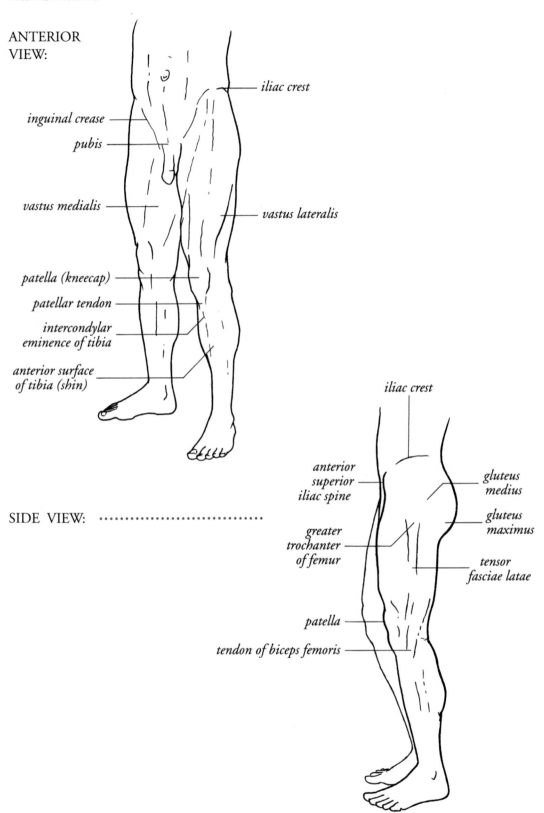

ANTERIOR
VIEW:

iliac crest

inguinal crease

pubis

vastus medialis

vastus lateralis

patella (kneecap)

patellar tendon

intercondylar
eminence of tibia

anterior surface
of tibia (shin)

SIDE VIEW:

iliac crest

anterior
superior
iliac spine

gluteus
medius

gluteus
maximus

greater
trochanter
of femur

tensor
fasciae latae

patella

tendon of biceps femoris

POSTERIOR VIEW:

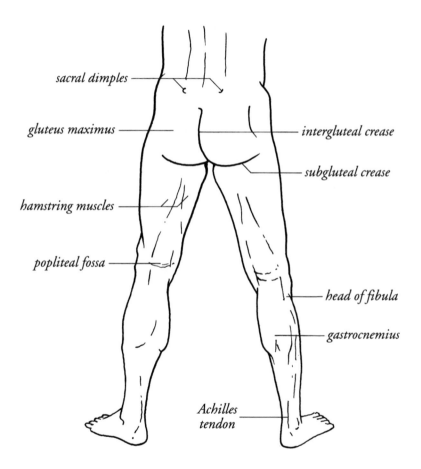

sacral dimples

gluteus maximus

intergluteal crease

subgluteal crease

hamstring muscles

popliteal fossa

head of fibula

gastrocnemius

Achilles
tendon

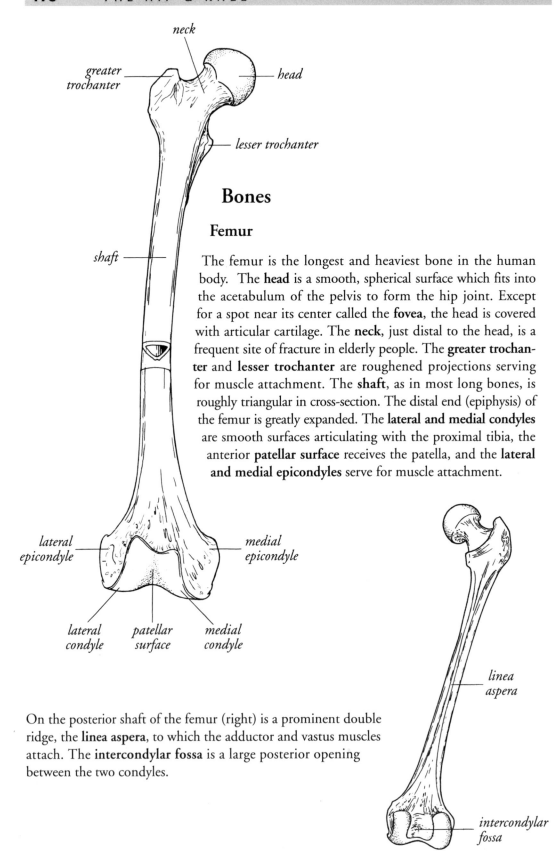

Bones

Femur

The femur is the longest and heaviest bone in the human body. The **head** is a smooth, spherical surface which fits into the acetabulum of the pelvis to form the hip joint. Except for a spot near its center called the **fovea**, the head is covered with articular cartilage. The **neck**, just distal to the head, is a frequent site of fracture in elderly people. The **greater trochanter** and **lesser trochanter** are roughened projections serving for muscle attachment. The **shaft**, as in most long bones, is roughly triangular in cross-section. The distal end (epiphysis) of the femur is greatly expanded. The **lateral and medial condyles** are smooth surfaces articulating with the proximal tibia, the anterior **patellar surface** receives the patella, and the **lateral and medial epicondyles** serve for muscle attachment.

On the posterior shaft of the femur (right) is a prominent double ridge, the **linea aspera**, to which the adductor and vastus muscles attach. The **intercondylar fossa** is a large posterior opening between the two condyles.

Patella

This is a sesamoid bone which develops within the tendon of the quadriceps muscle (right). Its broad superior edge is called the **base**, and the more pointed inferior edge is the **apex**.

base

The two **articular facets** on its posterior surface fit against the patellar surface of the femur, and are separated by a vertical ridge.

— *articular facets*

apex

Tibia and fibula

intercondylar eminence
lateral condyle
head of fibula
tibial tuberosity
anterior border
interosseous border

medial condyle

intercondylar eminence
lateral condyle

ANTERIOR VIEW   POSTERIOR VIEW

These are the two bones of the lower leg. [Note: by convention, we will hereafter refer to the lower leg as the "leg" and the upper leg as the "thigh."] The proximal tibia has **lateral and medial condyles** which fit against the identically-named surfaces of the distal femur. The medial tibial condyle is larger than the lateral one. The **intercondylar eminence** is a ridge separating the two condyles and fitting into the intercondylar fossa of the femur during flexion. There is a large **tibial tuberosity** on the anterosuperior tibia for attachment of the patellar ligament. There is a posterolateral facet below the lateral tibial condyle for articulation with the fibula. The lateral tibial shaft has a sharp **interosseous border** for attachment of the interosseous membrane which connects the tibia and fibula along the length of their shafts. The anterior tibial shaft, called the shin, is shaped like a sharp ridge. It lies just below the skin and is subject to frequent banging and bruising because of the lack of padding by muscles or adipose tissue.

The fibula lies parallel to the tibia and is considerably smaller. It does not participate in the knee joint, nor bear any weight. It serves for attachment of muscles and fasciae. The head of the fibula has a facet for articulation with the proximal tibia.

Hip joint

The acetabulum (Latin for "small bowl") is a deep socket formed by the junction of the ilium, pubis, and ischium (see page 41). It should not be confused with the obturator foramen (below).

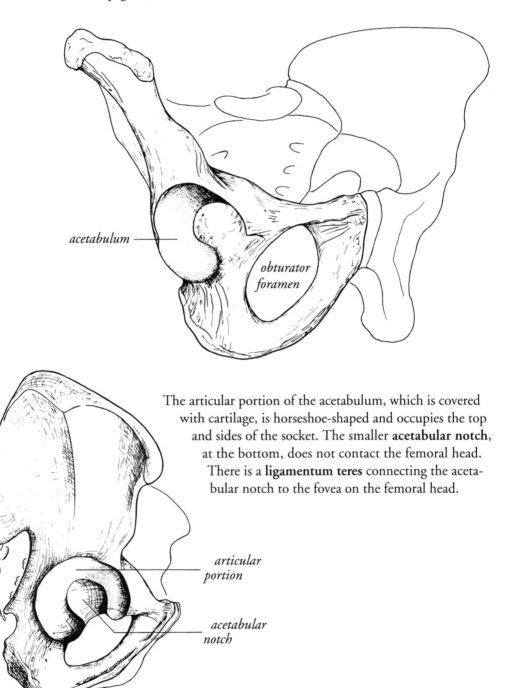

acetabulum

obturator foramen

The articular portion of the acetabulum, which is covered with cartilage, is horseshoe-shaped and occupies the top and sides of the socket. The smaller **acetabular notch**, at the bottom, does not contact the femoral head. There is a **ligamentum teres** connecting the acetabular notch to the fovea on the femoral head.

articular portion

acetabular notch

Because of the structure of the pelvis, the acetabulum is
directed laterally, anteriorly, and inferiorly. In cross section,
we see that the upper part of the socket, which fits against the
femoral head, is slightly oblique rather than horizontal (right).
This angle varies with the individual and with age. The more
the angle deviates from horizontal, the less stable is the joint.

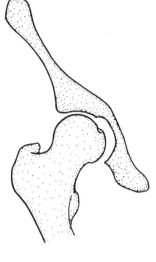

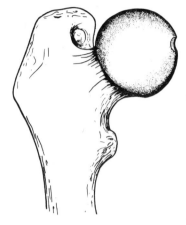

The femoral head represents
about ⅗ of a sphere and is
covered with thick cartilage,
except at the fovea where the
ligamentum teres is attached.

Looking at the head from the side and from above, we see that it is
oriented medially, superiorly, and anteriorly (right, and below). The
precise angle of the head, as well as the length of the femoral neck,
varies with the individual and with age.

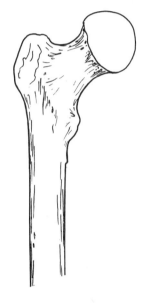

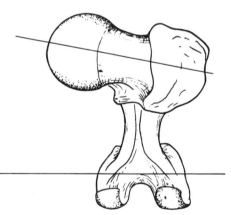

A fibrocartilaginous ring called the **labrum** is attached
around the rim of the acetabulum and is reinforced
by a **transverse acetabular ligament** which bridges the
inferior opening of the notch (right). The labrum helps
hold the femoral head in place, and increases the effective
depth of the socket.

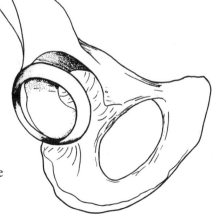

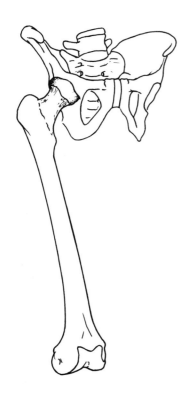

In anatomical position, the anterior femoral head is partly exposed.

The head fits better into the socket when the femur is flexed to a 90° angle relative to the trunk, as in a kneeling position.

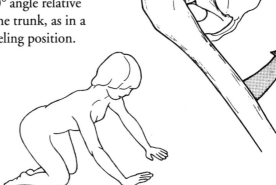

Maximal contact between the articular surfaces is attained by a combination of flexion, abduction and lateral rotation (left).

We often assume this position spontaneously when at rest.

Variations

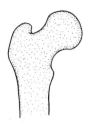

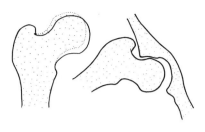

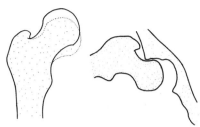

The average angle between the femoral neck and shaft is 135°.

In some individuals this angle is smaller, a condition called **coxa vara** in which the range of abduction is reduced.

When the angle is greater than 135° (**coxa valga**),the range of abduction is increased.

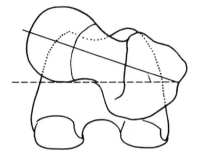

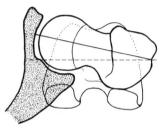

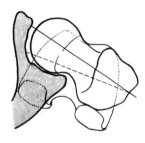

Seen from above, the neck is oriented anteriorly at an angle of 10 to 30°.

When this "anteversion" angle is small, the head fits into the socket well in anatomical position,

and maintains good articular contact even in lateral rotation.

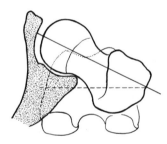

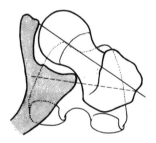

When the anteversion angle is large, the anterior part of the head is more exposed in anatomical position,

and the posterior part loses contact with the socket in lateral rotation.

Lateral rotation is more restricted in these individuals by contact between the neck and the lateral edge of the acetabulum. Curvature and length of the femoral neck also affect mobility at the hip joint.

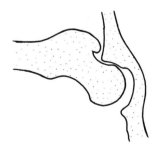

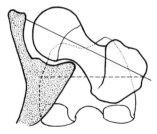

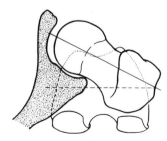

A neck that is more concave, and longer, will facilitate abduction...

and lateral rotation.

With a shorter, less concave neck, both these motions are restricted by contact with the edge of the acetabulum.

Obviously, there are intrinsic limitations to ROM at the hip joint due to the shape of the articulating bones, which may be greater or less in specific individuals. It is important to be aware of this variation when teaching dance or other physical disciplines. To some extent, restrictions of ROM at the hip joint can be compensated at nearby joints (e.g., lumbar spine, knee).

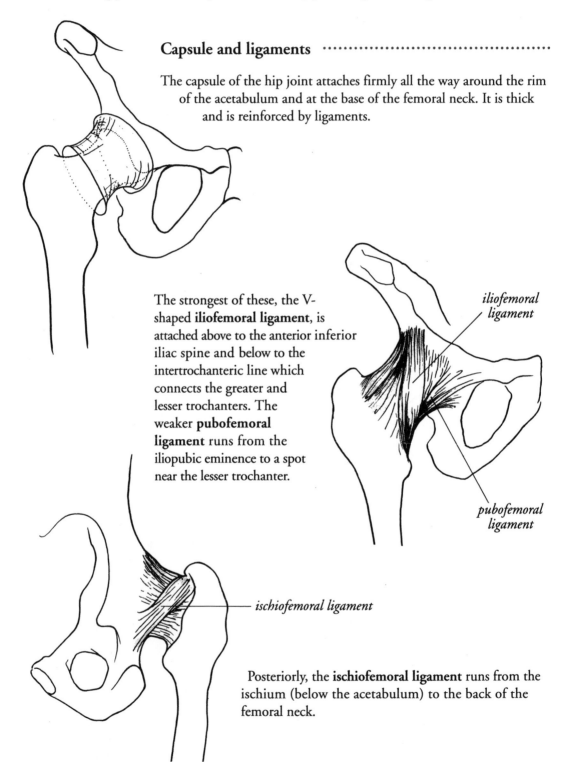

Capsule and ligaments ··

The capsule of the hip joint attaches firmly all the way around the rim of the acetabulum and at the base of the femoral neck. It is thick and is reinforced by ligaments.

The strongest of these, the V-shaped **iliofemoral ligament**, is attached above to the anterior inferior iliac spine and below to the intertrochanteric line which connects the greater and lesser trochanters. The weaker **pubofemoral ligament** runs from the iliopubic eminence to a spot near the lesser trochanter.

iliofemoral ligament

pubofemoral ligament

ischiofemoral ligament

Posteriorly, the **ischiofemoral ligament** runs from the ischium (below the acetabulum) to the back of the femoral neck.

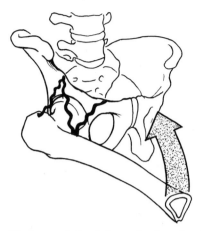

The iliofemoral (both branches) and pubofemoral ligaments become slack in flexion...

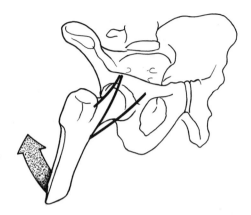

and taut in extension.

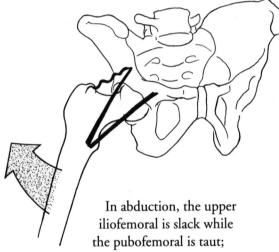

In abduction, the upper
iliofemoral is slack while
the pubofemoral is taut;

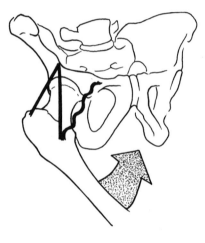

in adduction, the opposite occurs.

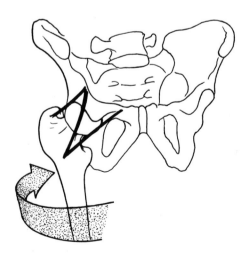

The ligaments all become taut in lateral rotation...

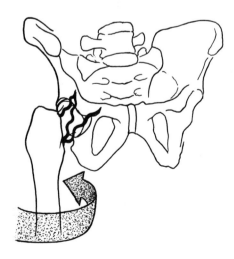

and slack in medial rotation.

Movements of hip

Some of these were mentioned above in relation to the bones and ligaments. We will now focus on the movements themselves. For simplicity, we assume first that the pelvis is fixed and the femur is moving.

Flexion is the movement in which the angle between the anterior surfaces of the thigh and the trunk decreases. ROM for hip flexion is greater when the knee is also flexed.

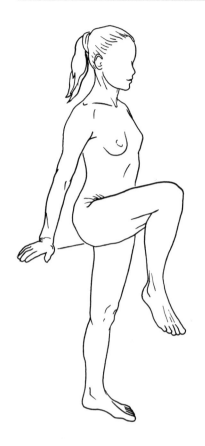

ROM for passive flexion is greater than for active flexion since the flexing muscles are relaxed and can be compressed.

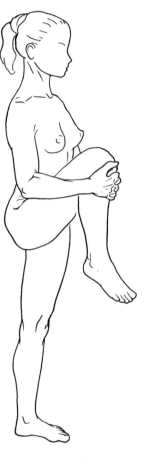

When the knee is extended, hip flexion is restricted by the limits of elasticity of the hamstring muscles (see page 221). Hip flexion is often associated with retroversion of the pelvis (see page 190).

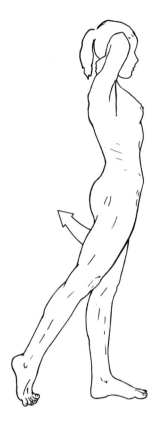

In **extension** of the hip, the angle between the posterior surfaces of the thigh and the trunk decreases (left). ROM for extension is limited compared to that for flexion, and this movement is often confused with or increased by lumbar lordosis (see page 31).

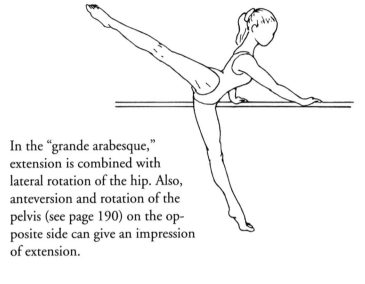

In the "grande arabesque," extension is combined with lateral rotation of the hip. Also, anteversion and rotation of the pelvis (see page 190) on the opposite side can give an impression of extension.

ROM for extension is reduced when the knee is flexed (left), because of the limits of elasticity of the rectus femoris muscle (see page 219).

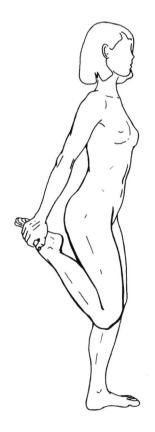

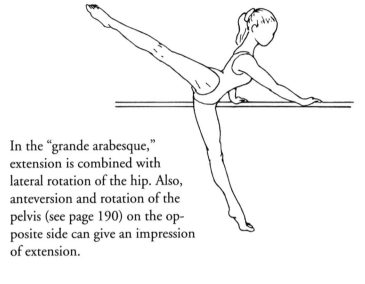

In **adduction** of the hip, the thigh moves towards or past the median plane. It can be combined with slight flexion (as shown here) or extension, with the other leg slightly displaced accordingly, so that the two legs can move past each other.

In **abduction** of the hip (left), the thigh moves away from the median plane, and the angle between the lateral surfaces of the thigh and the trunk decreases.

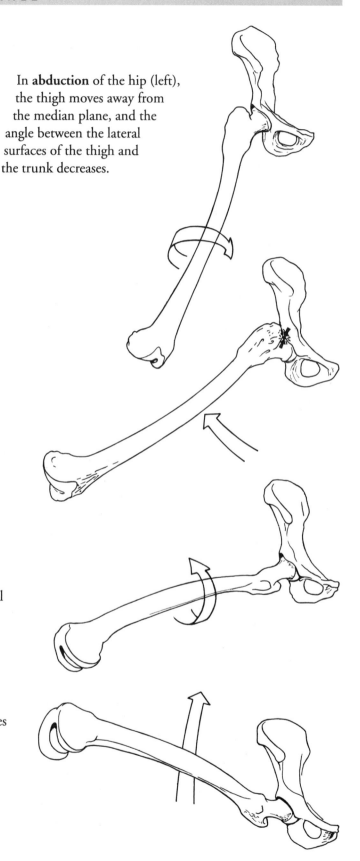

When the femur is in neutral or medial rotation (top right), abduction is limited to about 40° because of contact between the superior femoral neck and the upper edge of the acetabulum (right).

However, with the femur in lateral rotation,

the inferior aspect of the neck faces the edge of the socket, and ROM for abduction is greater.

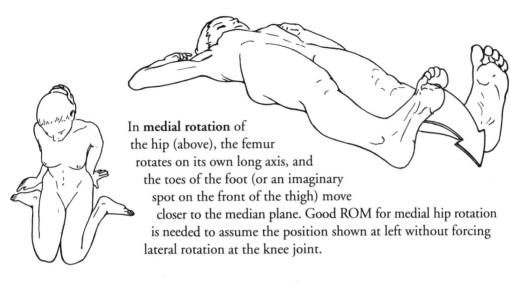

In **medial rotation** of the hip (above), the femur rotates on its own long axis, and the toes of the foot (or an imaginary spot on the front of the thigh) move closer to the median plane. Good ROM for medial hip rotation is needed to assume the position shown at left without forcing lateral rotation at the knee joint.

In **lateral rotation** of the hip (below), the toes (or a spot on the front of the thigh) move away from the median plane.

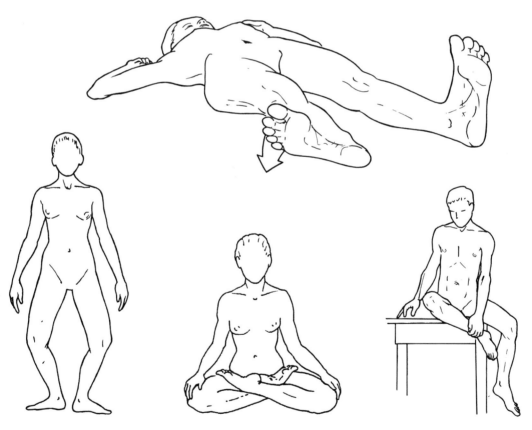

Good ROM for lateral rotation is needed for the "en dehors" position of ballet,

or for assuming the "lotus position" without stressing the knee and ankle joints.

When the hip is flexed, ROM for lateral rotation is greater because the ilio-femoral ligament is slack.

Let us now consider the possible movements *of the pelvis* at the hip joint, assuming that the femur is fixed. We will focus on the anterior superior iliac spine (ASIS) as a reference point.

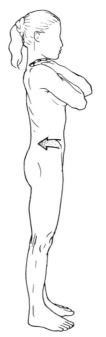

It can move forward (**ante-version** of the pelvis), which tends to increase lordosis of the lumbar spine.

The ASIS can move backward (**retroversion**), which decreases lumbar lordosis.

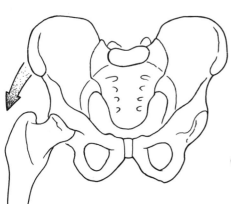

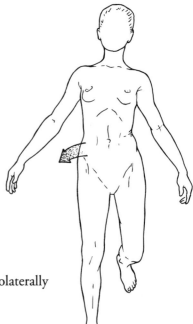

It can move inferolaterally (lateral flexion)...

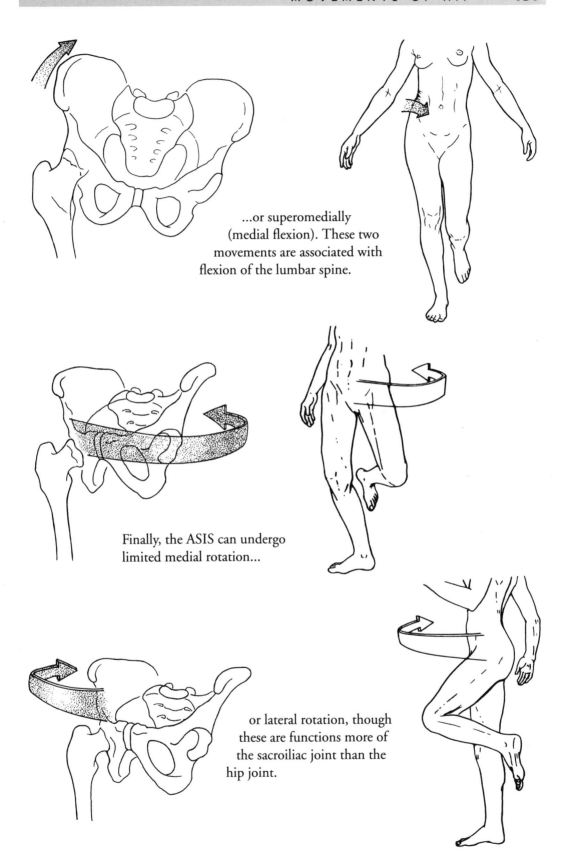

...or superomedially (medial flexion). These two movements are associated with flexion of the lumbar spine.

Finally, the ASIS can undergo limited medial rotation...

or lateral rotation, though these are functions more of the sacroiliac joint than the hip joint.

Knee joint

Three bones contribute to this joint. The femur articulates with both the tibia and patella; however, the tibia does not articulate with the patella (far right).

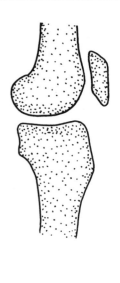

The distal femur and proximal tibia, where they come together, are both expanded, like the ends of two columns (right). This increases their stability and weight-bearing ability.

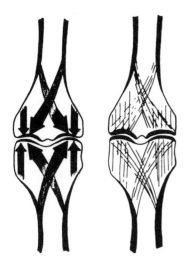

The fibers of the alveolar (spongy) tissue inside are oriented in diagonal as well as longitudinal directions, which increases strength (left). The lateral and medial condyles of the femur are convex and separated by a shallow groove anteriorly, which turns into the deep intercondylar fossa posteriorly (right). The overall shape is reminiscent of a rocking chair.

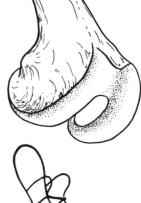

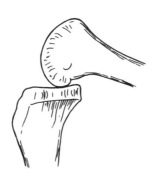

The condyles are less curved anteriorly (good for weight-bearing function, above left) and more curved posteriorly (good for flexion movement, above right). Overall, the medial condyle is more curved than the lateral condyle, which helps explain the automatic rotations of the knee during flexion/extension (see page 202). Prolonged standing with the knee in slightly flexed position (right) puts stress on a small articular surface of the condyles, and can damage the cartilage.

The lateral and medial condyles of the tibia are concave for articulation with the convex condyles of the femur (right). The raised intercondylar eminence fits into the groove and intercondylar fossa of the femur.

How do the femoral condyles move during flexion? If they were to simply roll backward like a ball rolling down a hill, the femur would slip off the tibia (below, left). If they were to glide in one spot like a ball bearing, a single spot on the tibia would receive all the friction, and the cartilage there would be damaged (below, right).

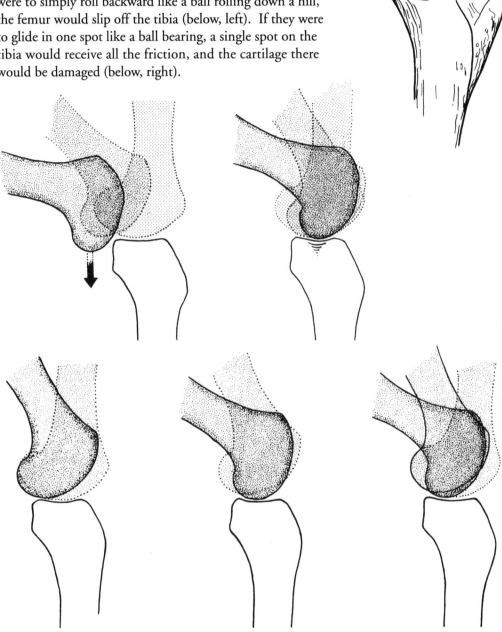

In fact, the femoral condyle first rolls (15-20°) on the tibial condyle,

then glides,

producing a combined "rolling-gliding" movement.

The opposite occurs in extension of the knee: first gliding, then rolling.

For the lower limb, we can consider three different axes.
The first ("mechanical axis") passes through the middle of
the femoral head above and the middle of the ankle joint
below (see below, left). In anatomical position, this axis is at
an angle of about 3° from a vertical (sagittal) plane (shown
as "V" in the diagram). If you stand on one foot, this axis
moves farther from the sagittal plane. The second and third
axes are those passing through the shafts of the femur and
tibia (see below, right).

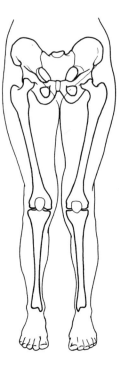

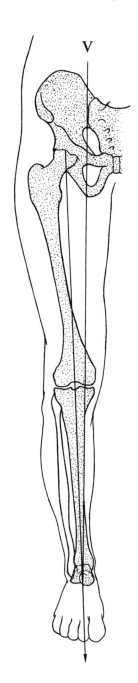

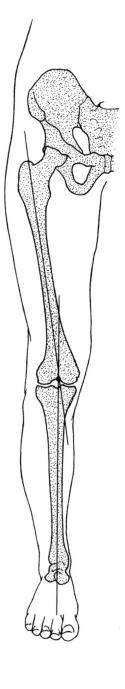

The lateral angle formed by
these two axes varies from
person to person; it can be
less than 180° ("genu valgum,"
above) or greater than 180°
("genu varum," below).

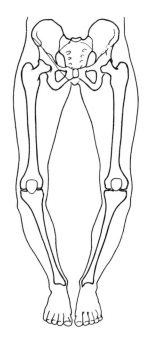

Menisci

The **menisci** (singular: meniscus) are C-shaped intra-articular discs made of fibrocartilage. Their shallow central tips are attached to the intercondylar eminence of the tibia, and their thicker margins are attached to the peripheral edges of the tibial condyles.

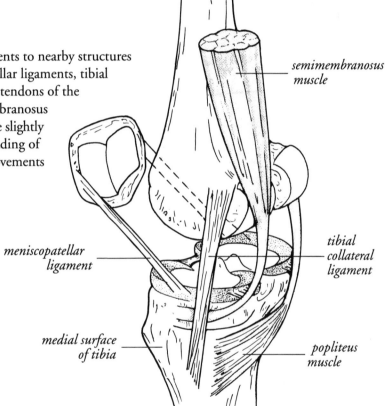

They also have attachments to nearby structures such as the meniscopatellar ligaments, tibial collateral ligament, and tendons of the popliteus and semimembranosus muscles. The menisci are slightly mobile, and aid in spreading of synovial fluid during movements of the knee.

semimembranosus muscle

meniscopatellar ligament

tibial collateral ligament

medial surface of tibia

popliteus muscle

By providing a greater surface for articulation with the femoral condyles, they also permit better weight-bearing function (distribution of forces) and better stability of the knee joint.

without meniscus   *with meniscus*

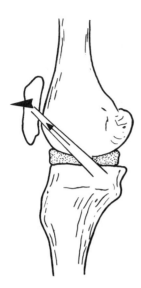

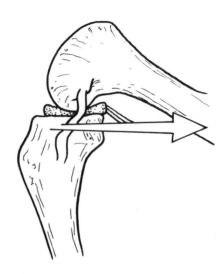

In extension, the menisci move forward because they are (i) pushed in that direction by the femoral condyles, and (ii) pulled by the meniscopatellar ligaments, which are in turn pulled upward by movement of the patella.

In flexion, the menisci move backward because they are (i) pushed in that direction by the condyles, and (ii) pulled by the tendons of the semimembranosus and popliteus muscles, and the tibial collateral ligament.

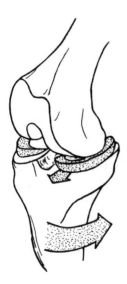

In rotation, the ipsilateral meniscus moves forward because of pressure from the condyle.

These movements are all necessary for normal function of the knee joint. In some cases (particularly rapid extension movements, as in soccer), the menisci may not move fast enough, and become crushed or torn.

Capsule and ligaments

The **capsule of the knee joint** attaches just outside the articular surfaces of the three bones involved. The patella is contained in the anterior capsule. Thus, the patella, femur, tibia, and capsule enclose a single synovial cavity within which synovial fluid circulates. The capsule is very slack anteriorly (left), which allows good ROM for flexion (middle). In extension, therefore, the capsule forms deep folds at the front and sides. In cases of prolonged immobilization of the joint, these folds can develop adhesions which subsequently limit flexion. In terms of bone morphology, and in comparison with the elbow, the knee is not a very stable or well-fitted joint. Therefore, its ligaments are essential for its stability.

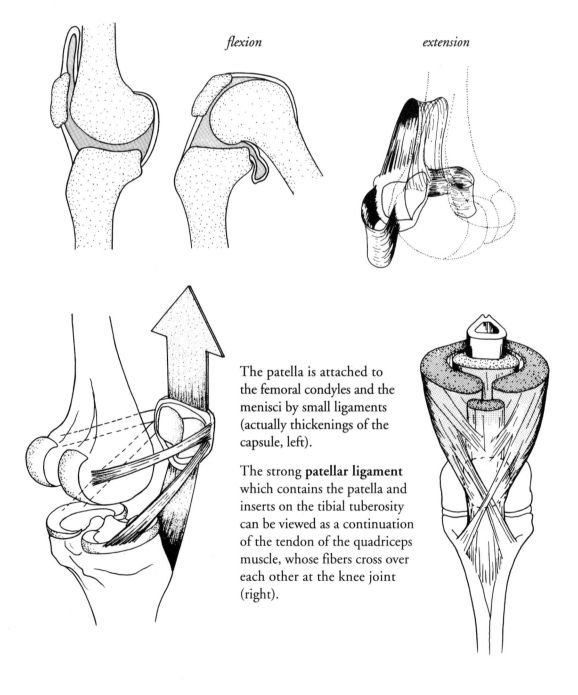

flexion   *extension*

The patella is attached to the femoral condyles and the menisci by small ligaments (actually thickenings of the capsule, left).

The strong **patellar ligament** which contains the patella and inserts on the tibial tuberosity can be viewed as a continuation of the tendon of the quadriceps muscle, whose fibers cross over each other at the knee joint (right).

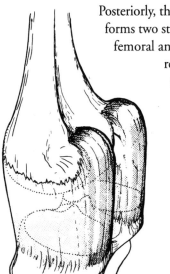

Posteriorly, the knee capsule is thicker and forms two strong bands connecting the femoral and tibial condyles. These resist hyperextension of the joint, and provide "passive stability" in the standing position (see page 201).

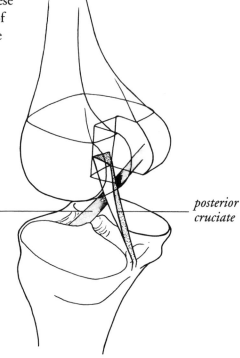

anterior cruciate —————— *posterior cruciate*

There are two cruciate ("crossed") ligaments located in the intercondylar fossa of the femur. They are named according to where they attach to the tibia. Anatomically, they are outside the joint capsule. The **anterior cruciate ligament** is attached to the anterior intercondylar area of the tibia. It runs posterosuperolaterally and attaches to the medial aspect of the lateral femoral condyle.

The **posterior cruciate ligament**, which is stronger than the anterior cruciate, attaches to the posterior intercondylar area of the tibia. It runs anterosuperomedially and attaches to the lateral surface of the medial femoral condyle. The anterior cruciate tends to resist anterior displacement of the tibia on the femur (far left), while the posterior cruciate resists posterior displacement (near left).

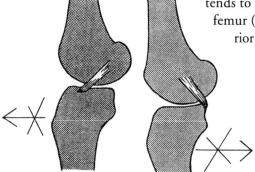

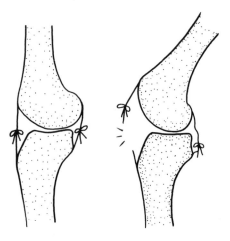

Why have obliquely-oriented ligaments perform this braking action? Because simple anterior and posterior ligaments would not allow flexion.

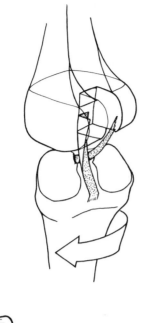

In both flexion and extension, the cruciate ligaments remain fairly taut, and displacement of the tibia on the femur is minimal. In lateral rotation, the cruciates slacken somewhat (left). In medial rotation, they press against each other, and therefore become more taut (right).

On the sides, the joint capsule is reinforced by two collateral ligaments. The **medial (tibial) collateral ligament** runs from the medial epicondyle of the femur to the medial condyle and upper medial shaft of the tibia (left). Its lower attachment is slightly anterior relative to its upper attachment. The medial collateral functions to stabilize the joint and prevent it from opening on the medial side. If this ligament is ruptured, the tibia will be able to move laterally (right).

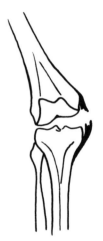

The **lateral (fibular) collateral ligament** runs from the lateral epicondyle of the femur to the head of the fibula (left). Its lower attachment is slightly posterior relative to its upper attachment. This ligament prevents the joint from opening on the lateral side. If it ruptures, the tibia will be able to move medially (right).

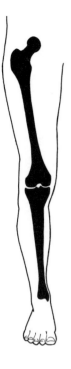

The medial collateral is considerably thicker and stronger than its lateral counterpart. Why? In the average person, the lateral angle formed by the femur and tibia is slightly less than 180° ("genu valgum," see page 194). Since the joint "gapes" more on the medial side, there is a need for stronger stabilization on that side.

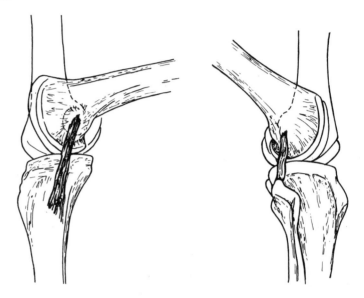

The collaterals tend to be taut in extension (left) and slack in flexion (right). Therefore, they resist hyperextension.

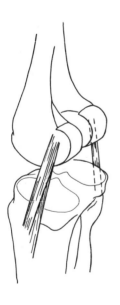
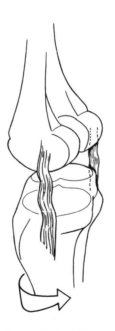
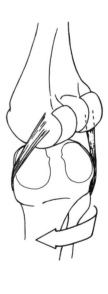

On leg bones which have been "pulled apart" for illustrative purposes (left), we see that the collaterals become slack in medial rotation of the tibia (middle) due to their orientation, and taut in lateral rotation (right). Therefore, they resist excessive lateral rotation.

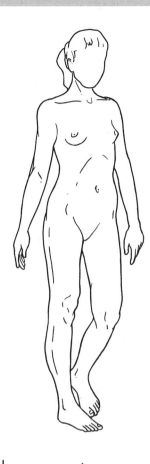

The knee ligaments act together to stabilize the joint. In extension, all the ligaments are taut, and the joint can be passively stabilized without any muscular action, e.g., when balancing on one foot.

Here (left), the joint is "locked" in slight hyperextension by the tautness of the ligaments (particularly the thickened posterior portion of the joint capsule).

In flexion (right), most of the ligaments are slack, and the joint therefore has some capacity for rotation.

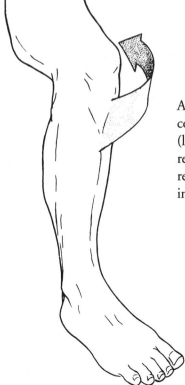

As noted above, the cruciates and collaterals tend to limit medial (left) and lateral (right) rotation, respectively. But they are more restrictive in the extended than in the flexed position.

Some "automatic" rotation of the knee occurs during flexion/extension. Why is this? The primary explanation involves the shape of the femoral and tibial condyles. The medial femoral condyle is more curved than the lateral one, i.e., its radius of curvature is smaller.

medial

lateral

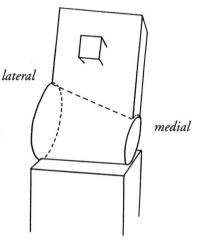

To understand the implications of this, visualize the two condyles as fitting inside a truncated cone, and the femoral shaft as a rectangular slab with a projection which we shall use as a reference landmark (right).

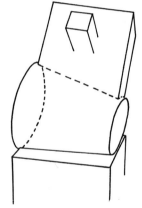

During flexion, due to the shape of the cone, the landmark becomes directed somewhat laterally (left).

The tibial condyles are also not totally symmetric (right); both are concave transversely, but from front to back the lateral condyle is slightly convex while the medial one is concave. Therefore, the lateral tibial condyle allows more rolling than does the medial one.

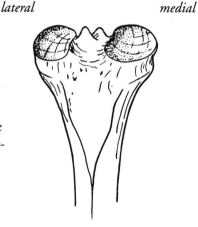

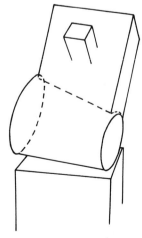

During flexion, the lateral femoral condyle rolls backward more than does the medial one, which accentuates the lateral orientation of our landmark, i.e., the lateral rotation of the femur.

The secondary explanation for automatic rotation of the knee is that the medial collateral ligament is stronger than the lateral one, as we mentioned above. This reinforces the tendency of the medial femoral condyle to be less mobile than the lateral one.

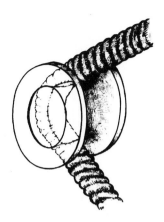

Functions of patella

What exactly does the patella do? Well, obviously it protects the knee joint itself from external impact, e.g., falling forward onto your knees. It also protects the quadriceps tendon, in which it is contained. During movements, this tendon slides in the groove between the femoral condyles, like a rope in a pulley.

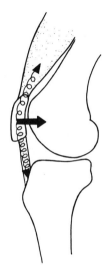

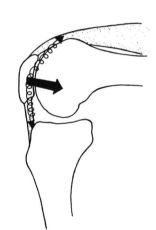

The patella is pressed against the groove during flexion; this pressure can be 400 kg or more during squatting.

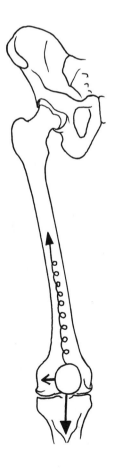

The patella is not stable laterally. The quadriceps follows the femoral shaft and its force is slightly oblique, but its tendon runs straight down to insert on the tibia. Thus, contraction of the quadriceps tends to pull the patella laterally (left), just as a pulley would move sideways if its rope came down at an angle (right). This instability is maximal during active extension or slight flexion when the patella is not firmly pressed against the patellar surface of the femur; during full flexion the patella is better "locked" in place.

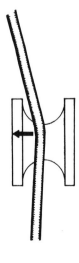

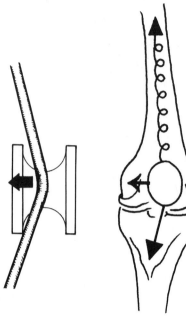

The instability of the patella is accentuated during lateral rotation of the tibia when the lower as well as upper part of the tendon becomes obliquely oriented.

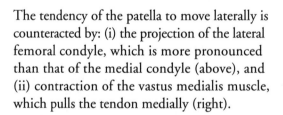

The tendency of the patella to move laterally is counteracted by: (i) the projection of the lateral femoral condyle, which is more pronounced than that of the medial condyle (above), and (ii) contraction of the vastus medialis muscle, which pulls the tendon medially (right).

As you can see, the articulation of the patella against the femur is subjected to major strains and stresses, particularly on the lateral side. This explains the frequency of arthritis here, which can compromise proper gliding of the patella and active extension of the knee.

Movements of knee

The knee is primarily a hinge joint, although it does undergo some automatic rotation as explained above.

Flexion decreases the angle formed by the posterior thigh and leg (right).

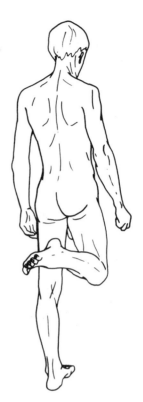

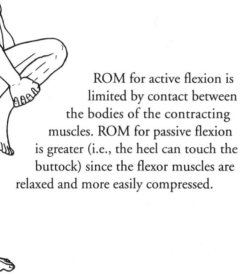

ROM for active flexion is limited by contact between the bodies of the contracting muscles. ROM for passive flexion is greater (i.e., the heel can touch the buttock) since the flexor muscles are relaxed and more easily compressed.

In addition, ROM for flexion is greater when the hip joint is flexed and smaller when the hip is extended. Why? Because position at the hip joint affects the degree of tension in the rectus femoris muscle (see page 219).

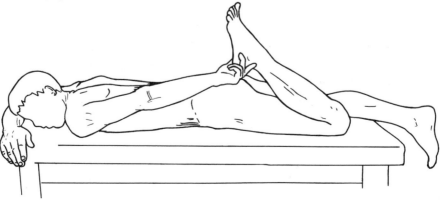

Extension is an increase in the angle between the posterior thigh and leg, i.e., a return from flexion back to anatomical position.

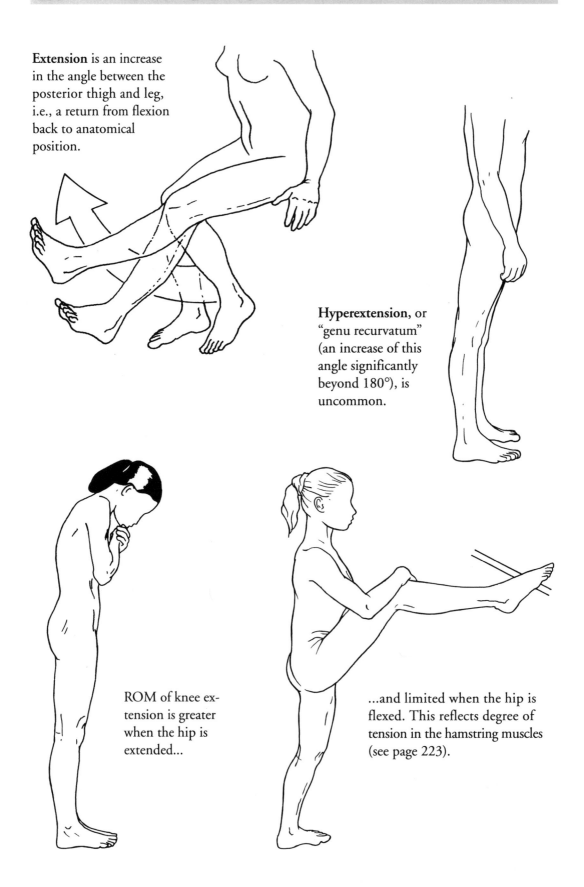

Hyperextension, or "genu recurvatum" (an increase of this angle significantly beyond 180°), is uncommon.

ROM of knee extension is greater when the hip is extended...

...and limited when the hip is flexed. This reflects degree of tension in the hamstring muscles (see page 223).

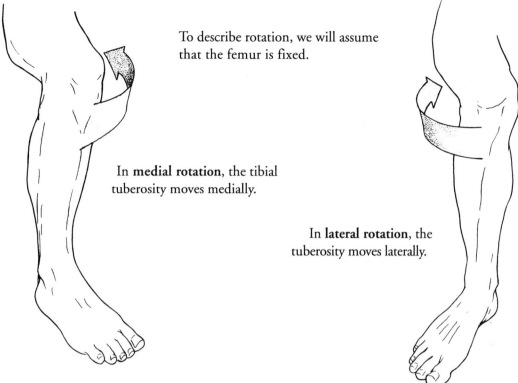

To describe rotation, we will assume that the femur is fixed.

In **medial rotation**, the tibial tuberosity moves medially.

In **lateral rotation**, the tuberosity moves laterally.

Rotation can occur to an appreciable extent only when the knee is flexed, and the ligaments are relaxed (see page 201). It occurs automatically during knee flexion and is due primarily to the shape of the condyles (see page 202). If the knee is extended and you see the tuberosity moving medially or laterally, this is rotation not at the knee but at the hip (right).

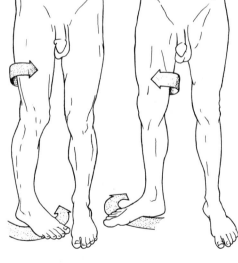

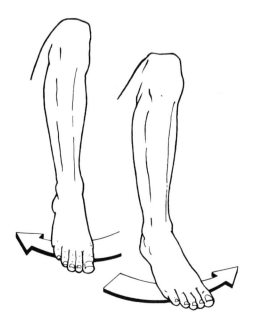

It is also important not to confuse knee rotation with abduction/adduction of the foot. This is the reason for focusing on movement of the tibial tuberosity rather than the foot.

Muscles of hip

A group of six deep hip muscles can be seen in this superior view of the pelvis.

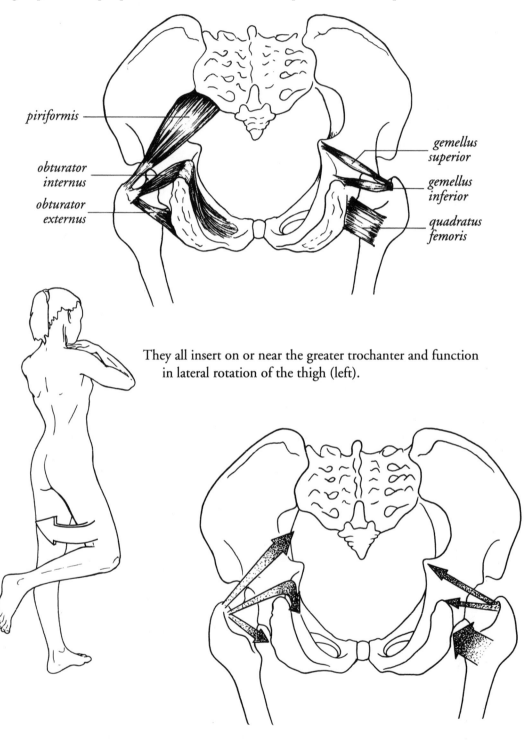

piriformis

obturator
internus

obturator
externus

gemellus
superior

gemellus
inferior

quadratus
femoris

They all insert on or near the greater trochanter and function
in lateral rotation of the thigh (left).

The forces produced by their individual
contractions are illustrated above.

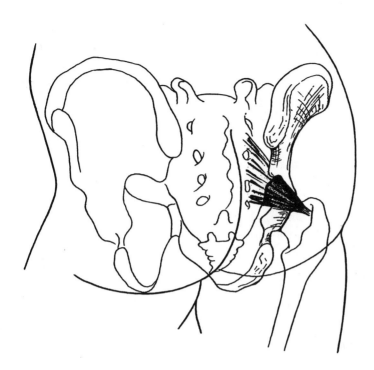

Piriformis originates on the anterior sacrum (segments 2-4), passes under the greater sciatic notch, and inserts on the top of the greater trochanter.

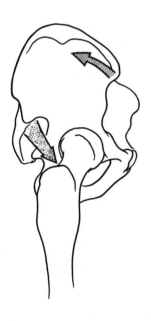

If the sacrum is fixed, piriformis laterally rotates the extended thigh or abducts the flexed thigh. If the femur is fixed, it contributes to retroversion of the pelvis (on bilateral contraction, left), or to medial rotation of the pelvis (on unilateral contraction, right).

[Movements of the pelvis are defined on pages 190-191.]

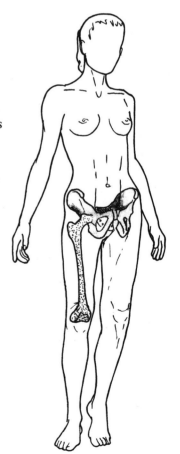

Quadratus femoris runs from the lateral ischium to the posterior aspect of the greater trochanter (left). Its actions, both on the thigh and the pelvis, are the same as those of piriformis (below).

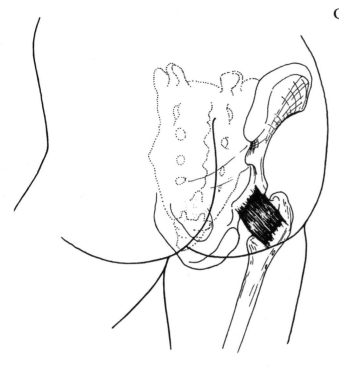

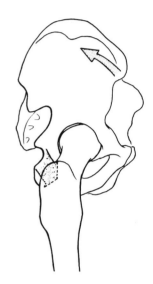

Obturator internus arises from the obturator membrane and adjacent portions of the ischium and ilium. Its fibers pass posteriorly through the lesser sciatic notch, make a sharp bend around the body of the ischium, and insert on the medial aspect of the greater trochanter (below right). There is a bursa where it wraps around the ischium, to reduce friction. This muscle laterally rotates the thigh and helps stabilize the hip joint because of its broad origin.

If the femur is fixed, the muscle acts in retroversion (below), medial rotation, or medial flexion of the pelvis.

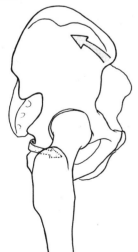

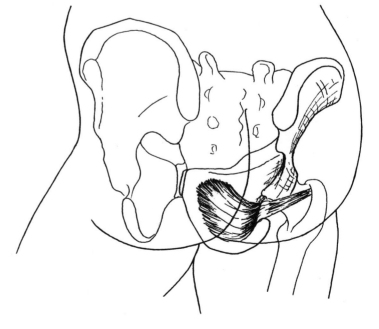

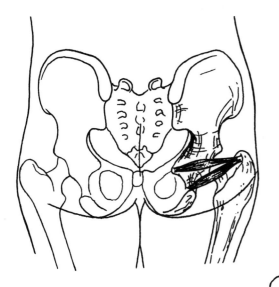

Gemellus superior and inferior are small muscles located above and below the distal borders of obturator internus, whose actions they reinforce.

Obturator externus arises from the external surface of the obturator membrane, passes posterior to the femoral neck, and inserts on a fossa on the medial surface of the greater trochanter. Its position makes it the ideal lateral rotator of the thigh.

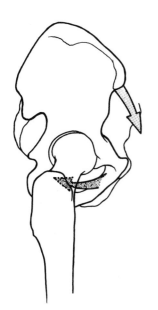

If the femur is fixed, it functions in anteversion (left), medial rotation, or medial flexion (right) of the pelvis.

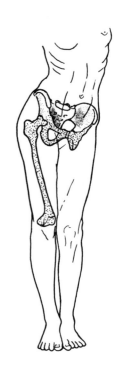

If we look at the hip from the side, we observe that obturator internus and the gemelli run from the greater trochanter in a posteroinferior direction, while obturator externus runs anteroinferiorly. The combined action of the obturators and gemelli, therefore, can be viewed as follows:

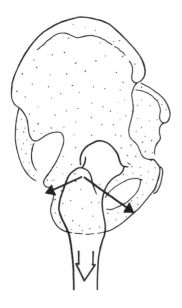

If the pelvis is fixed, they will pull the femur down relative to the pelvis.

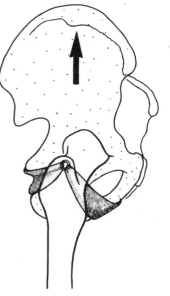

If the femur is fixed, they will lift the pelvis relative to the femur.

Either way, they tend to "pull apart" the hip joint, on a very small scale. This is a decompressive effect which is quite beneficial for certain painful conditions (e.g., worn-down cartilage).

The obturators and gemelli have been compared to a "hammock" supporting the pelvis.

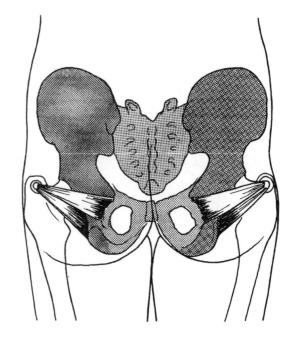

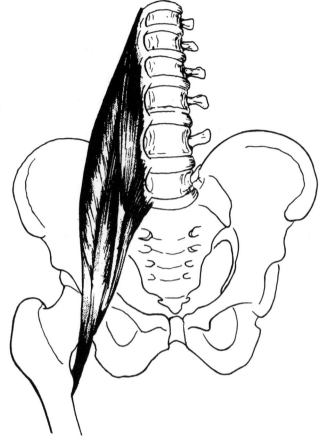

Psoas major arises from the bodies of T12 through L5, and from arches of fascia which connect the boney parts of the vertebral bodies but do not attach to the intervertebral disks.

It runs anterior to the pelvis, posterior to the inguinal ligament, and inserts on the lesser trochanter.

There is a bursa to reduce friction where it bends at the anterior pelvis.

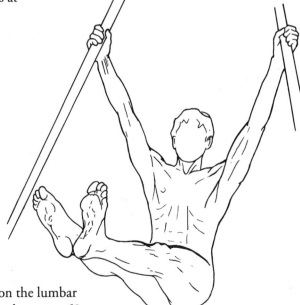

Psoas is the primary hip flexor. Its effect on the lumbar spine when the femur is fixed was described on page 62.

Iliacus arises from the entire internal iliac fossa. Inferiorly, its fibers merge with those of psoas and insert on the lesser trochanter via the same tendon.

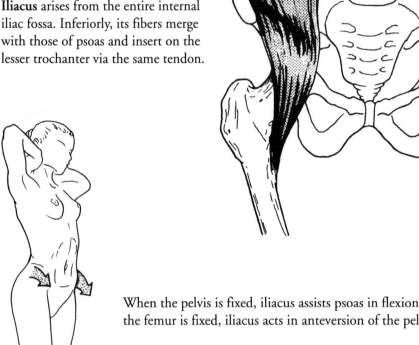

When the pelvis is fixed, iliacus assists psoas in flexion of the hip. When the femur is fixed, iliacus acts in anteversion of the pelvis (left).

Because they share the same tendon and have the same action on the thigh, iliacus and psoas are often described as a single muscle ("iliopsoas"). However, it is important to remember that their superior attachments are quite different. When the femur is fixed, iliacus acts on the pelvis, whereas psoas acts on the lumbar spine.

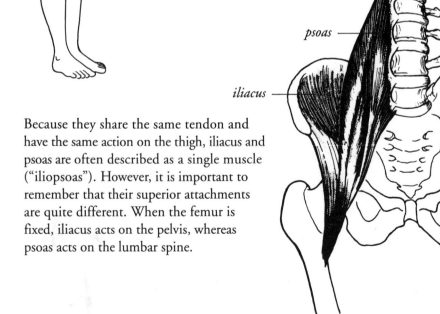

psoas

iliacus

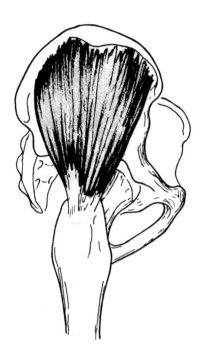

Gluteus medius has a broad origin on the external iliac fossa. Its fibers converge and insert on the lateral aspect of the greater trochanter.

Its major action is abduction of the hip (right), but it can also assist in flexion and extension. When the femur is fixed, gluteus medius is involved in both anteversion and retroversion of the pelvis, depending on whether the anterior or posterior fibers contract (bilateral contraction).

With unilateral contraction, it acts in lateral flexion of the pelvis, and also stabilizes the pelvis during walking or standing on one foot.

Gluteus minimus is a small muscle originating just anterior to gluteus medius and inserting on the anterior aspect of the greater trochanter. Its actions reinforce those of the anterior part of gluteus medius.

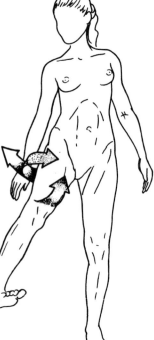

In addition to abduction of the thigh, it assists in flexion and medial rotation.

If the femur is fixed, gluteus minimus assists in anteversion (left)...

...lateral flexion (right), or rotation of the pelvis.

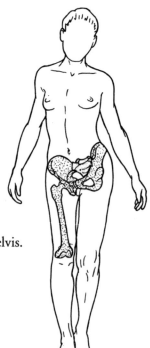

Quadriceps femoris is a massive muscle having four bodies (rectus femoris, vastus lateralis, vastus medialis, and vastus intermedius) which converge into a single quadriceps tendon. This tendon inserts on and surrounds the patella, then continues as the patellar ligament to insert on the tibial tuberosity.

Vastus inter-medius arises on the upper two thirds of the anterior femoral shaft.

Vastus lateralis and medialis arise from either side of the linea aspera on the posterior femoral shaft (left), then wrap around the sides to meet anteriorly, superficial to vastus intermedius (right).

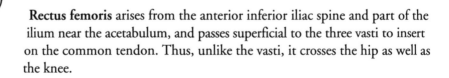

Rectus femoris arises from the anterior inferior iliac spine and part of the ilium near the acetabulum, and passes superficial to the three vasti to insert on the common tendon. Thus, unlike the vasti, it crosses the hip as well as the knee.

With its four bodies acting together, the quadriceps extends the knee. It is the strongest muscle in the body.

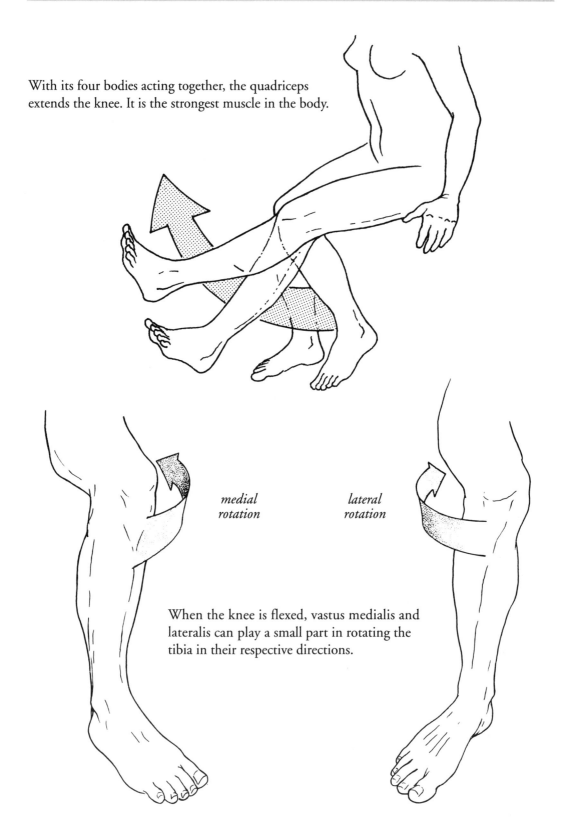

medial rotation   *lateral rotation*

When the knee is flexed, vastus medialis and lateralis can play a small part in rotating the tibia in their respective directions.

When the knee is extended, these two muscles act as stabilizers of the patella and knee joint.

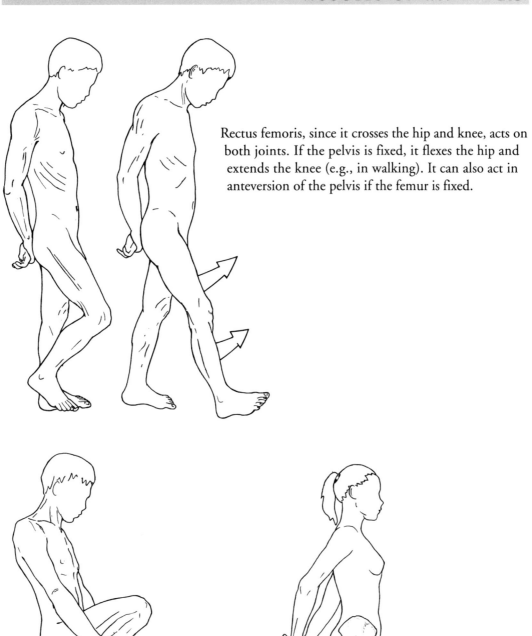

Rectus femoris, since it crosses the hip and knee, acts on both joints. If the pelvis is fixed, it flexes the hip and extends the knee (e.g., in walking). It can also act in anteversion of the pelvis if the femur is fixed.

The three vasti can be stretched by full flexion of the knee and hip.

For stretching rectus femoris, the hip must be in extension and the knee in flexion.

Sartorius is the longest muscle in the body. It originates from the ASIS, runs medially down the thigh, superficial to quadriceps, and inserts on the superomedial shaft of the tibia. Its name means "tailor's muscle" and refers to the cross-legged position used by tailors in the old days.

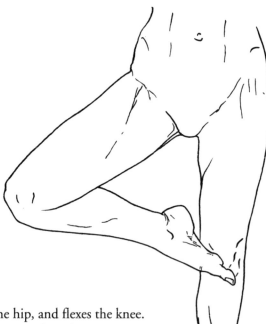

Sartorius flexes, laterally rotates, and abducts the hip, and flexes the knee. If the femur is fixed, it anteverts and medially rotates the pelvis.

The **hamstrings** are a group of three posterior muscles working together to flex the knee and extend the thigh.

Semimembranosus arises from the ischial tuberosity and inserts on the posteromedial aspect of the tibial condyle (left).

Semitendinosus has a similar origin, and inserts via a very long, thin tendon to the superomedial tibial shaft (right). Its tendon blends with those of sartorius and gracilis.

Biceps femoris has two heads. The long head arises from the ischial tuberosity, at which point it cannot be distinguished from semitendinosus. The short head (which is partially covered by the long head) arises from the posterior femoral shaft. The two heads merge inferiorly and insert via a common tendon to the head of the fibula (left). This tendon is bifurcated by the lateral collateral ligament of the knee.

Collectively, the hamstrings originate from the same place but spread out to have both medial and lateral insertions on the proximal leg bones (right). Like rectus femoris and sartorius, the hamstrings are polyarticular muscles, i.e., they cross and act on more than one joint.

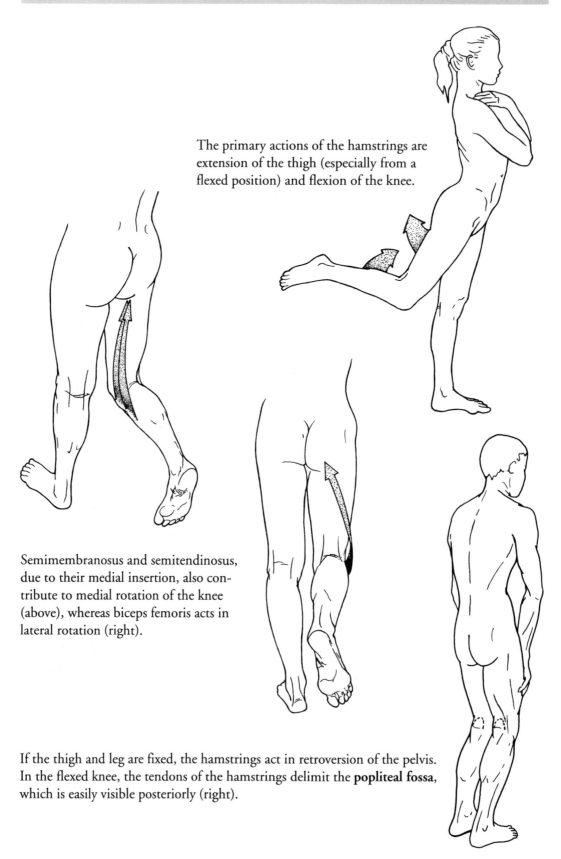

The primary actions of the hamstrings are extension of the thigh (especially from a flexed position) and flexion of the knee.

Semimembranosus and semitendinosus, due to their medial insertion, also contribute to medial rotation of the knee (above), whereas biceps femoris acts in lateral rotation (right).

If the thigh and leg are fixed, the hamstrings act in retroversion of the pelvis. In the flexed knee, the tendons of the hamstrings delimit the **popliteal fossa**, which is easily visible posteriorly (right).

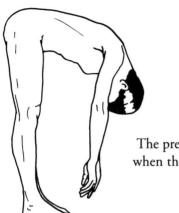

The presence of the hamstrings restricts ROM of hip flexion when the knee is extended (e.g., in toe touches).

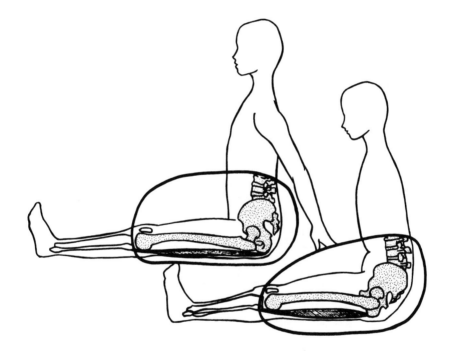

In a sitting position, with the knees extended, it can be difficult to sit directly on the ischial tuberosities, since the hamstrings tend to pull the pelvis into retroversion and thereby straighten out the lordosis (curvature) of the lumbar spine. Thus, hamstrings that are too tight can lead to problems with the spine, especially for dancers. Stretching warm-up exercises for the hamstrings are important in dance and most sports.

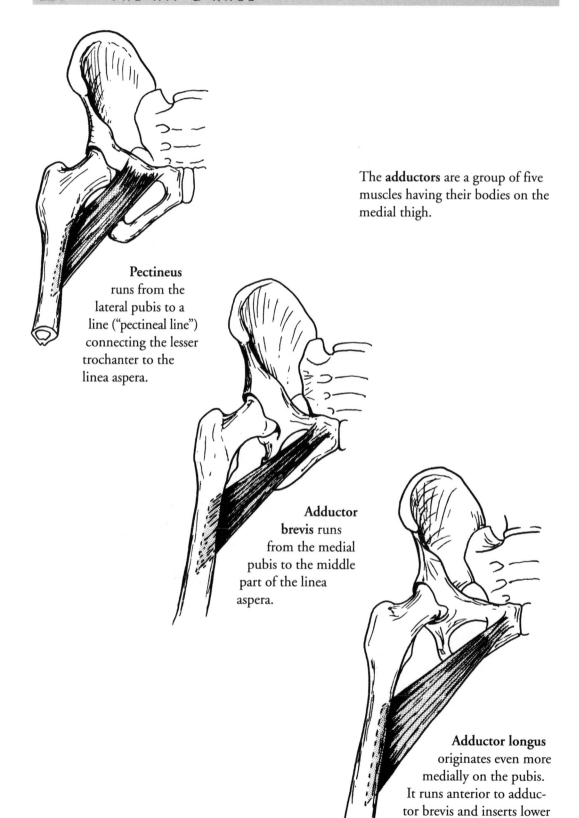

The **adductors** are a group of five muscles having their bodies on the medial thigh.

Pectineus runs from the lateral pubis to a line ("pectineal line") connecting the lesser trochanter to the linea aspera.

Adductor brevis runs from the medial pubis to the middle part of the linea aspera.

Adductor longus originates even more medially on the pubis. It runs anterior to adductor brevis and inserts lower on the linea aspera.

Gracilis is a long, thin, superficial, comparatively weak muscle running from the inferomedial pubis to a spot on the tibial shaft just below the medial condyle.

Adductor magnus, the largest and strongest of the adductor group, is really a compound muscle innervated by two different spinal nerves. Its anterior portion originates from the ischiopubic ramus, runs inferomedially, and has a very broad insertion on the linea aspera.

The posterior portion originates from the ischial tuberosity, runs straight down, and inserts just above the medial femoral condyle.

The gap between the two insertions is called the adductor hiatus.

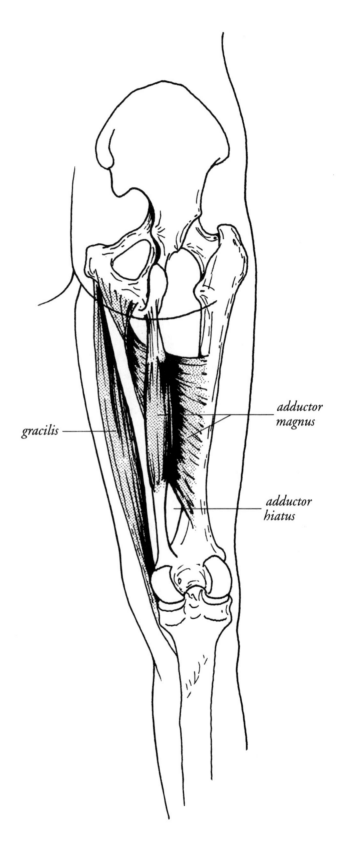

gracilis

adductor magnus

adductor hiatus

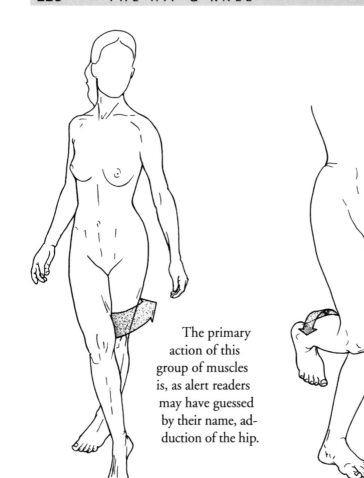

The primary action of this group of muscles is, as alert readers may have guessed by their name, adduction of the hip.

To a lesser degree, they can act from anatomical position as hip flexors or lateral or medial rotators. If the hip is in flexed position, they act as extensors. Gracilis, which is polyarticular, can also flex and medially rotate the knee.

If the femur is fixed, the adductors are involved in anteversion, medial flexion, lateral rotation, or (in the case of gracilis and the posterior portion of adductor magnus) medial rotation of the pelvis.

These muscles are frequently strained or torn ("pulled groin") during movements involving sudden or extreme abduction of the thigh.

Again, proper stretching exercises can help prevent this.

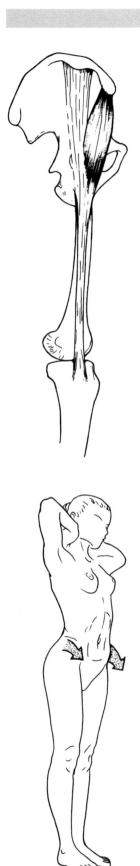

Tensor fasciae latae originates from the anterior iliac crest (near the ASIS) and runs inferiorly and slightly posteriorly. It inserts not on a bone, but rather on a band of strong fibrous tissue, called the fascia lata or iliotibial tract, which runs down the lateral thigh and attaches to the superolateral tibia and head of the fibula.

This muscle abducts, flexes, and medially rotates the thigh (right).

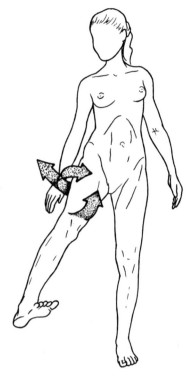

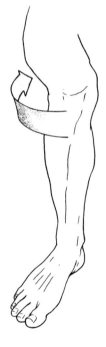

It plays a small part in knee extension or lateral rotation of the flexed knee.

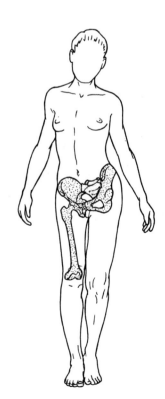

If the thigh and leg are fixed, it acts in anteversion (left), lateral flexion, or lateral rotation (right) of the pelvis.

Gluteus maximus is the largest muscle of the body and, together with considerable adipose tissue, provides the bulk of the buttocks.

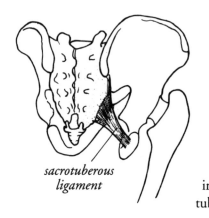

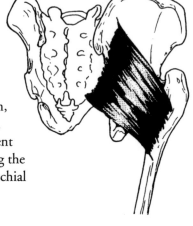

It has a broad origin which includes the posterolateral sacrum, posterior ilium, and sacrotuberous ligament (a ligament connecting the inferior sacrum to the ischial tuberosity, left).

sacrotuberous ligament

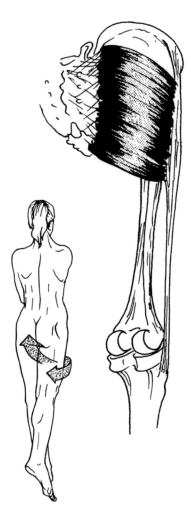

A deep portion of the gluteus maximus inserts on a roughened area ("gluteal tuberosity") of the postero-superior femoral shaft (above).

A larger superficial portion inserts on the superior part of the iliotibial tract (left).

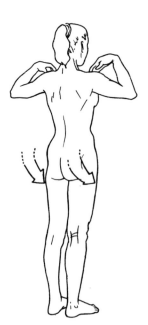

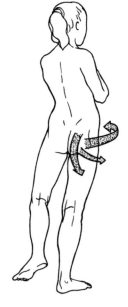

This muscle is the major hip extensor, and also acts in lateral rotation.

If the femur is fixed, it acts in retroversion,

medial rotation, and medial flexion of the pelvis.

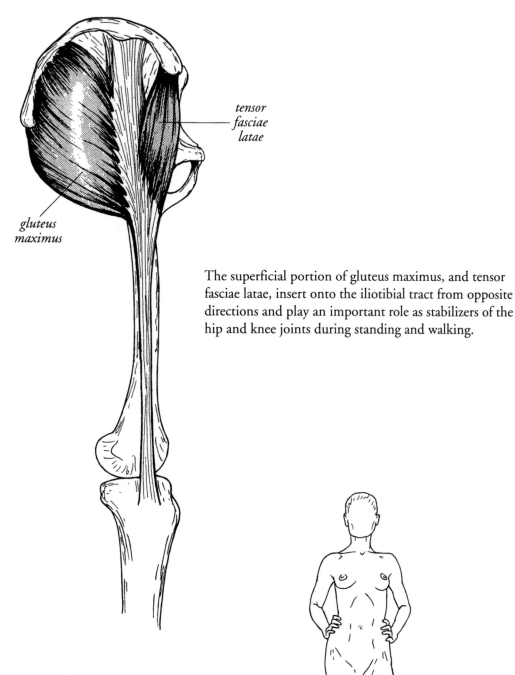

tensor
fasciae
latae

gluteus
maximus

The superficial portion of gluteus maximus, and tensor fasciae latae, insert onto the iliotibial tract from opposite directions and play an important role as stabilizers of the hip and knee joints during standing and walking.

They assist gluteus medius in maintaining the position of the contralateral pelvis while standing on one foot (see page 215).

Muscles of knee

Most muscles acting on the knee also act on the hip, and they have been described above. We need to mention only three muscles which do not cross the hip.

biceps femoris

The short head of **biceps femoris** arises from the femoral shaft and inserts on the head of the fibula. It flexes and laterally rotates the knee (see pages 221-222).

popliteus

Popliteus originates from the lateral aspect of the lateral femoral condyle, runs inferomedially, and inserts on a triangular area of the postero-superomedial tibial shaft. It flexes and medially rotates the knee (right).

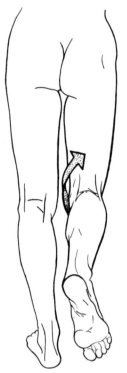

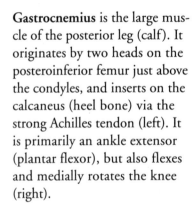

Gastrocnemius is the large muscle of the posterior leg (calf). It originates by two heads on the posteroinferior femur just above the condyles, and inserts on the calcaneus (heel bone) via the strong Achilles tendon (left). It is primarily an ankle extensor (plantar flexor), but also flexes and medially rotates the knee (right).

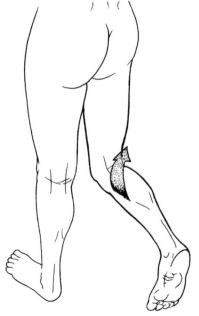

Summary of movements

We have covered many muscles and movements. Let us summarize the muscles involved in the specific movements of the hip and knee. The arrows represent the forces produced by the various muscles.

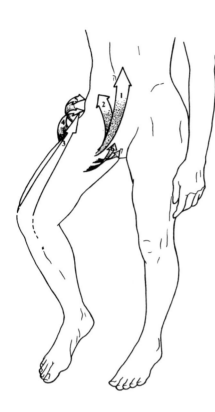

Flexion of hip (left):

(1) psoas
(2) iliacus
(3) rectus femoris
(4) tensor fasciae latae
(5) gluteus minimus and
    medius (anterior part)
(6) sartorius
(7) pectineus
not shown:
    gracilis

Extension of hip (right):

(1) gluteus maximus
(2) biceps femoris (long head)
(3) semimembranosus
(4) semitendinosus
(5) gluteus medius
    (posterior part)
not shown:
    adductor magnus

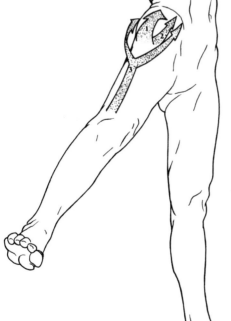

Abduction of hip (left):

(1) gluteus medius
(2) gluteus minimus
(3) tensor fasciae latae, gluteus maximus (superficial part)
not shown:
    piriformis, obturators, gemelli, sartorius

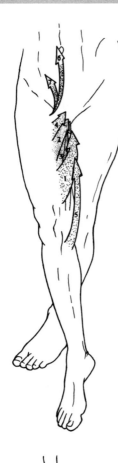

Adduction of hip (left):

(1) adductor magnus
(2) adductor longus
(3) adductor brevis
(4) pectineus
(5) gracilis
(6) psoas
(7) iliacus
not shown:
    biceps femoris (long head),
    gluteus maximus (deep part)

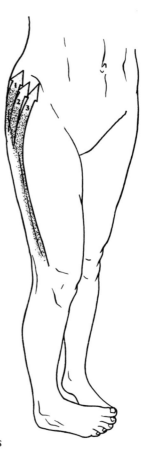

Medial rotation of hip (right):

(1) gluteus medius
(2) gluteus minimus
(3) tensor fasciae latae

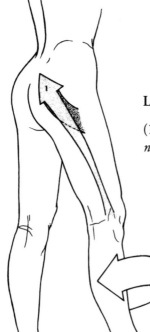

Lateral rotation of hip (left):

(1) gluteus maximus
not shown:
    piriformis, obturators, gemelli,
    quadratus femoris, biceps femoris
    (long head), adductors

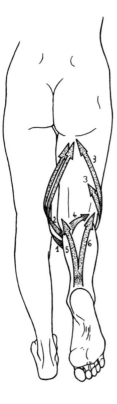

Flexion of knee (left):

(1) semitendinosus
(2) semimembranosus
(3) biceps femoris
(4) popliteus
(5, 6) gastrocnemius
not shown:
    sartorius, gracilis

Extension of knee (below):

(1) quadriceps femoris
(2) tensor fasciae latae,
    gluteus maximus
    (superficial part)

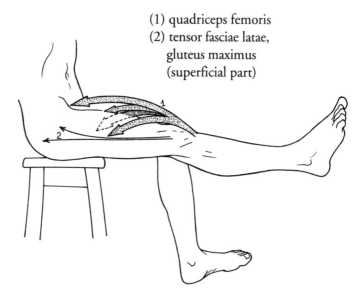

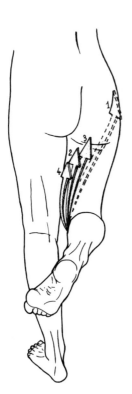

Medial rotation of knee (left):

(1) sartorius
(2) semitendinosus
(3) semimembranosus
(4) gracilis
not shown:
    popliteus

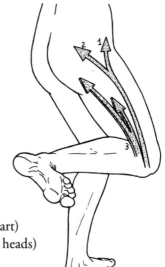

Lateral rotation of knee (right):

(1) tensor fasciae latae
(2) gluteus maximus (superficial part)
(3) biceps femoris (long and short heads)

Walking

Let us examine the sequential actions of the muscles during walking.

The leg, free from the body's weight, does an oscillating movement which takes the foot forward.

Contraction of rectus femoris (flexes hip, then extends knee), completed by contraction of entire quadriceps (extends knee).

The body's weight shifts onto the leg again.

Contraction of lateral stabilizing muscles of hip and knee.

Contraction of quadriceps, hamstrings, gemelli, gluteus maximus.

The leg propels the body forward.

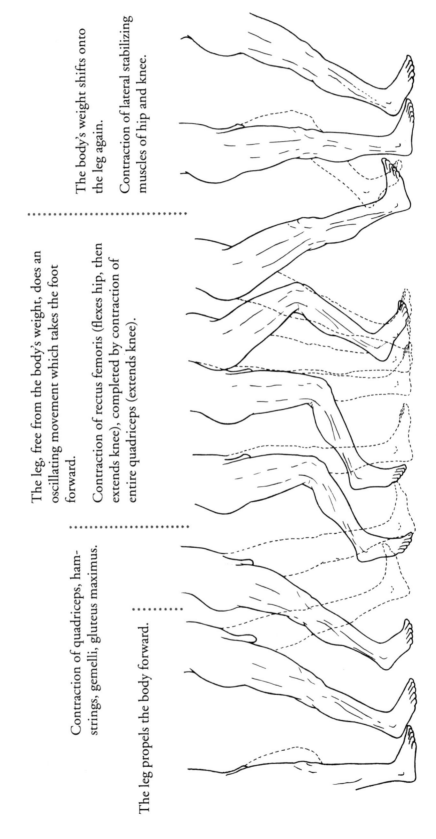

CHAPTER SEVEN

The Ankle & Foot

....

In most mammals, the weight of the body is distributed among four feet. As humans adopted a bipedal posture during evolution, the foot took on a difficult double function: bearing the weight of the entire body, and performing the complex movements necessary for walking, running, etc. This requires both strength and flexibility. The foot contains 26 bones, 31 joints, and 20 intrinsic muscles! Unfortunately, in today's world, the foot is subjected to many unusual stresses (e.g., hard pavements, poorly-fitting shoes, high heels) which may result in malfunction or even malformation. By understanding its structure and function, we are better able to avoid injury.

In this chapter, we will describe both the foot and the ankle joint by which it is connected to the leg.

Landmarks

ANTERIOR VIEW:

MEDIAL VIEW:

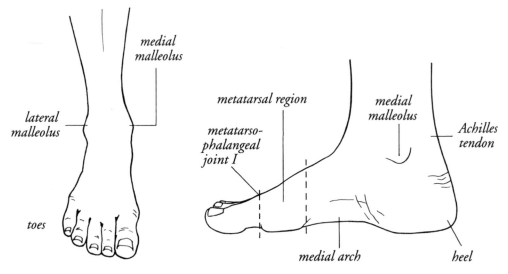

medial malleolus

lateral malleolus

toes

metatarsal region

metatarso-phalangeal joint I

medial malleolus

Achilles tendon

medial arch

heel

POSTERIOR VIEW:

INFERIOR VIEW:

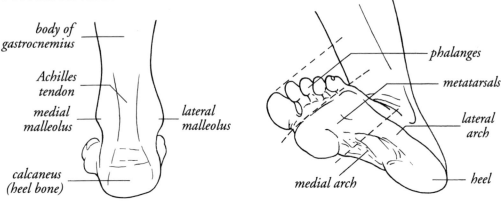

body of gastrocnemius

Achilles tendon

medial malleolus

calcaneus (heel bone)

lateral malleolus

phalanges

metatarsals

lateral arch

heel

medial arch

LATERAL VIEW:

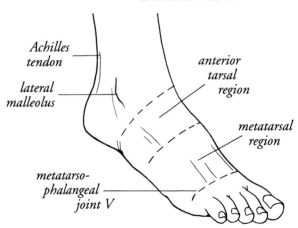

Achilles tendon

lateral malleolus

metatarso-phalangeal joint V

anterior tarsal region

metatarsal region

As you can see on a wet footprint, the medial arch and proximal phalanges do not normally come in contact with the ground. The heel, lateral arch, distal metatarsal region, and distal phalanges do contact the ground.

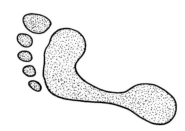

Bones

Tibia and fibula

The proximal ends of these two leg bones were described in Chapter 6. Distally, the tibia and fibula form a **medial malleolus** and **lateral malleolus**, respectively, to "clasp" the talus (the uppermost tarsal bone) and increase the stability of the ankle joint.

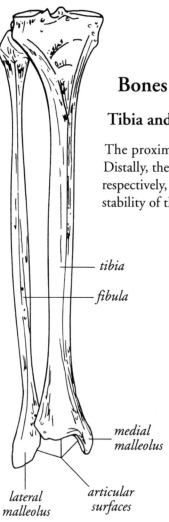

tibia

fibula

medial malleolus

lateral malleolus

articular surfaces

The distal end of the tibia, and inner aspects of the two malleoli, feature smooth articular surfaces to fit against corresponding surfaces of the talus.

tibia

fibula

proximal tibiofibular joint

The **proximal tibiofibular joint** is a true synovial joint, with a synovial cavity and a capsule, reinforced by anterior and posterior ligaments. It allows limited movement. Along the length of their shafts, the tibia and fibula are connected by the **interosseous membrane**. The **distal tibiofibular joint** is a syndesmosis (fibrous joint) in which the two bones are connected by ligaments and fibrous tissue, but without a synovial cavity or capsule.

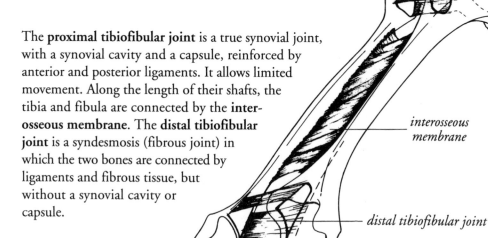

interosseous membrane

distal tibiofibular joint

Talus and calcaneus

These are the most posterior and massive of the tarsal bones. **Talus** articulates with the tibia and fibula above and with **calcaneus** below. Both articulate anteriorly with other tarsals. Interestingly, no muscle inserts on talus. It is moved indirectly via the structures surrounding it.

The posterosuperior and anteroinferior parts of talus are called the body and head respectively. They are connected by a thinner neck. The body has three articular surfaces for articulation with the corresponding surfaces of the tibia and fibula (following page.) The head articulates anteriorly with the navicular bone and inferiorly with calcaneus.

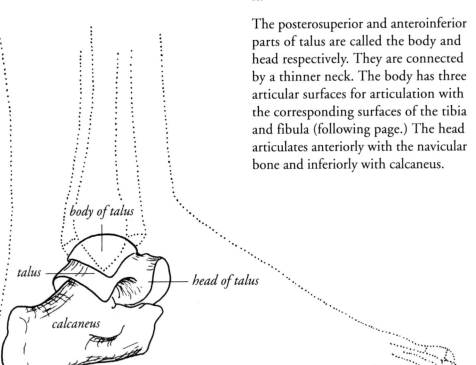

body of talus

talus

head of talus

calcaneus

Calcaneus is a large, irregularly-shaped bone. There
is a posteroinferior tuberosity for contact with the
ground. Superiorly, near the middle, is a surface for
articulation with talus. The anterior projection articulates superiorly with talus and anteriorly with the
cuboid. A thick layer of fibrous tissue called the **plantar
aponeurosis** runs from the inferior calcaneus along
the plantar surface of the foot, finally splitting into
five slips which attach to the toes. This aponeurosis
is thickest in the middle part, where it contributes to
the arches of the foot.

LATERAL VIEW:

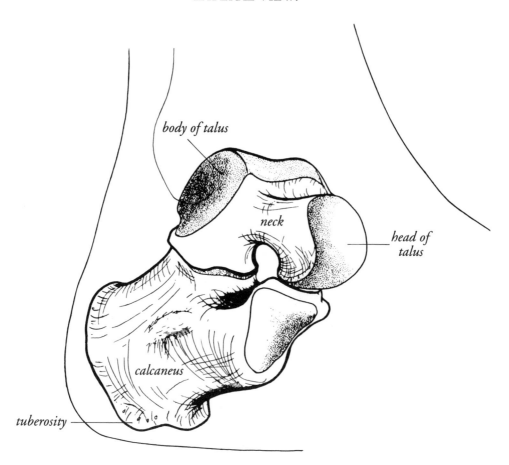

Medially, we see two tubercles on the body of talus, with a groove between them for passage of the flexor hallucis longus tendon. On calcaneus, we see a prominent projection called the **sustentaculum tali** (this helps support talus), a groove for passage or various tendons, blood vessels and nerves, and a large insertion area for the Achilles tendon.

MEDIAL VIEW:

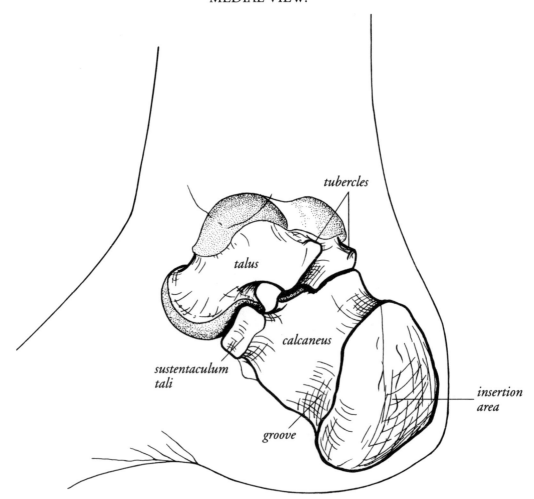

tubercles

talus

calcaneus

sustentaculum tali

insertion area

groove

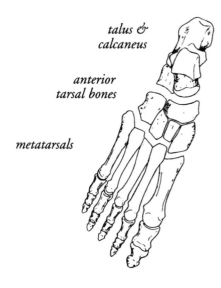

*talus &
calcaneus*

*anterior
tarsal bones*

metatarsals

Anterior tarsal bones

Between the talus and calcaneus, and the metatarsals, are five small bones collectively called the anterior tarsals (left).

They correspond to the portion of the external foot called the "instep."

instep

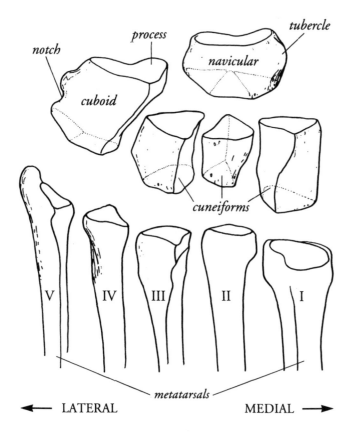

notch

process

tubercle

cuboid

navicular

cuneiforms

V   IV   III   II   I

metatarsals

◄— LATERAL   MEDIAL —►

Navicular articulates with talus. It is concave on the proximal end and convex on the distal end, where there are three facets for articulation with the cuneiforms. Medially, it has an externally palpable tubercle for insertion of tibialis posterior.

Cuboid does not actually bear much resemblance to a cube. It has facets for articulation with calcaneus (proximally), lateral cuneiform and navicular (medially), and metatarsals IV and V (distally). A proximal process fits under calcaneus and helps maintain the lateral arch of the foot. A lateral notch continues as a groove on the inferior surface and accommodates the tendon of peroneus longus.

The **cuneiforms** are three small wedge-shaped bones with the sharp edges directed inferiorly. They articulate proximally with navicular and distally with metatarsals I-III. Together with cuboid and the metatarsals, they constitute the transverse arch of the foot.

The anterior tarsals, and their many gliding joints, allow a reasonable degree of flexibility, though less than that of the corresponding wrist bones.

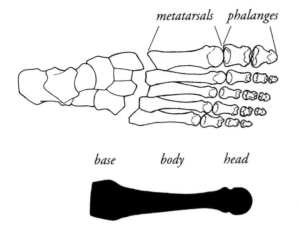

metatarsals *phalanges*

Metatarsals and phalanges

The structure of these bones is similar to that of the corresponding metacarpals and phalanges of the hand, except that the big toe, unlike the thumb, is not opposable.

base *body* *head*

Each **metatarsal** consists of a proximal base, a body, and a distal head.

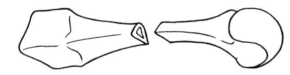

The base is roughly quadrangular, with facets for articulation with the tarsals and adjacent metatarsals. The head is convexly rounded, with a cartilaginous surface for articulation with the proximal phalanx, and a tiny tubercle on each side.

The body is triangular in cross-section, like most long bones.

The proximal **phalanx** of each toe has a concavely rounded base for articulation with the metatarsal, and a pulley-shaped head. The base of the middle phalanx is concave but with a median crest to match the shape of the head of the proximal phalanx. The head of the distal phalanx is flared to support the toenail superiorly, and has an inferior tubercle to support the fleshy part of the toe.

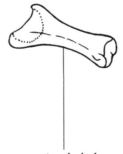

proximal phalanx *middle phalanx* *distal phalanx*

Arrangement of bones

We can view the foot as consisting of five
"spokes," i.e., metatarsals and toes I-V and
the tarsal bones behind them.

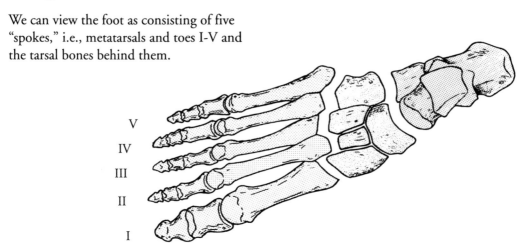

The "lateral foot" consists of calcaneus,
cuboid, and metatarsals/phalanges IV-V.
It includes the lateral arch and is more
involved in receiving and supporting the
body's weight during walking or running.

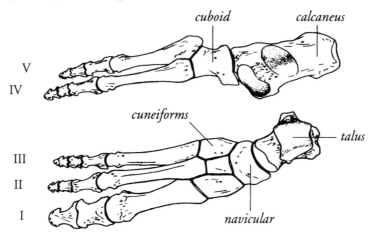

The "medial foot" consists of talus, navicular,
cuneiforms, and metatarsals/phalanges I-III.
It includes the medial arch and is more involved in propulsion, i.e., initiating the next
step during walking.

Joints

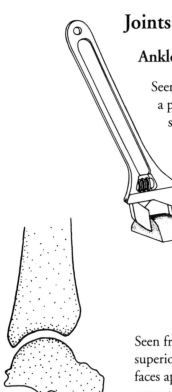

Ankle joint

Seen from the front, this joint resembles a pincer or monkey wrench gripping a section of a hemisphere.

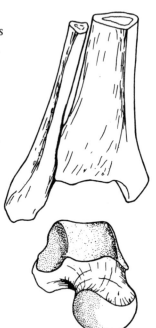

The lateral and medial malleoli, and distal tibia, fit against the three facets of the talar body (right).

Seen from the side, the cartilaginous superior and inferior articulating surfaces appear concave and convex, respectively (left).

SUPERIOR VIEW:

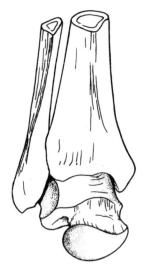

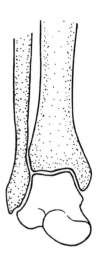

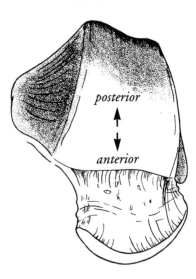

posterior

anterior

The two malleoli provide a snug fit around the talus. The lateral malleolus extends farther down and is more obliquely oriented than the medial malleolus.

In cross-section, we see that there is a slight ridge on the articulating surface of the distal tibia, and corresponding groove on talus.

We should also note that the superior talus is wider anteriorly than posteriorly.

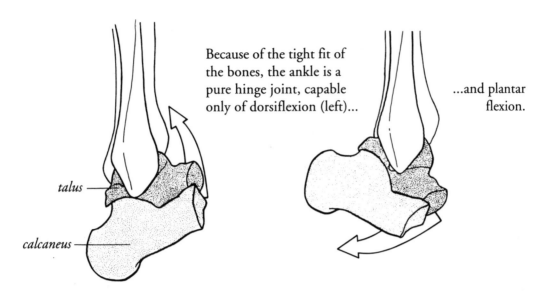

Because of the tight fit of the bones, the ankle is a pure hinge joint, capable only of dorsiflexion (left)...

...and plantar flexion.

talus

calcaneus

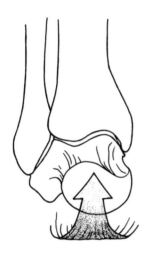

Of all the joints in the foot, this one allows the greatest range of motion (ROM). In dorsiflexion, the anterior (wider) part of the superior talus moves into the "pincer," and the joint is more stable (left). In plantar flexion, the posterior (narrower) part of the talus is in the "pincer," and the joint therefore less stable (right). This lack of stability is compensated in part by support from surrounding muscles and ligaments.

The **anterior and posterior talofibular ligaments** run from the lateral malleolus to the front and back ends of talus, while the **calcaneofibular ligament** runs down to the lateral calcaneus.

talofibular ligaments:

posterior

anterior

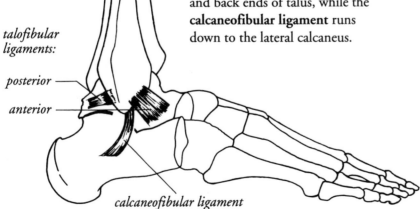

calcaneofibular ligament

Medially, there is a large **deltoid ligament** running from the medial malleolus to talus, sustentaculum tali, calcaneonavicular ligament, and navicular tuberosity.

Deep to this are two smaller ligaments (anterior and posterior) running from the medial malleolus to talus (below).

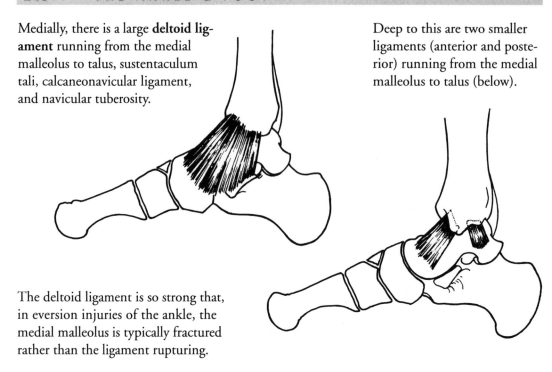

The deltoid ligament is so strong that, in eversion injuries of the ankle, the medial malleolus is typically fractured rather than the ligament rupturing.

In dorsiflexion, the posterior ligaments are taut, and the anterior ones are slack (left).

The reverse is true in plantar flexion. Since the ankle is least stable in plantar flexion, a "sprained ankle" occurs most commonly in this position, and the anterior talofibular ligament is the one most often injured.

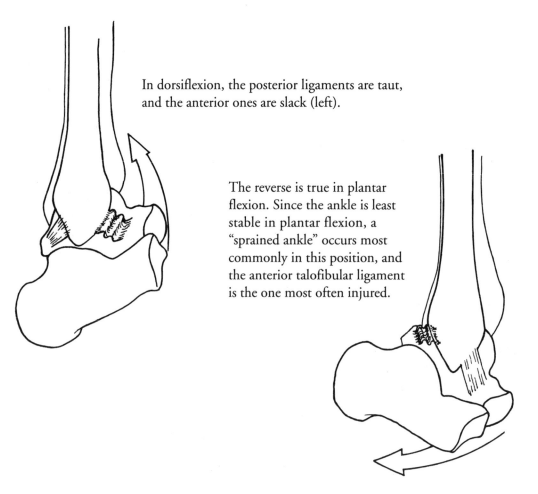

Subtalar (talocalcaneal) joint

Talus sits slightly obliquely on calcaneus, since the long axes of the two bones are oriented (respectively) somewhat medially and laterally (right).

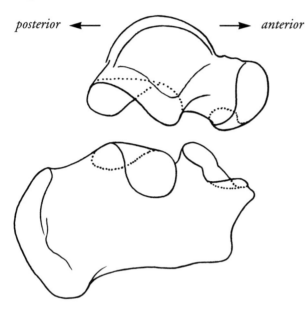

posterior ←   → *anterior*

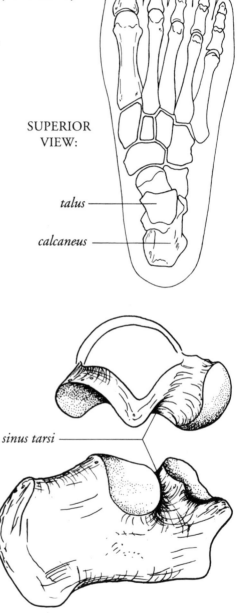

SUPERIOR VIEW:

talus

calcaneus

Anteriorly, a convex surface of talus fits against a concave surface of calcaneus, while posteriorly a concave surface of talus fits against a convex surface of calcaneus (above).

sinus tarsi

The **sinus tarsi** is a groove between the two bones which contains the strong **interosseous talocalcaneal ligament** and blood vessels.

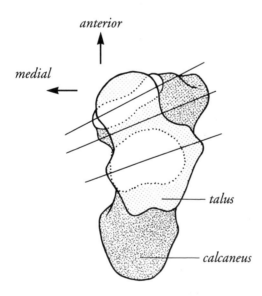

anterior

medial

talus

calcaneus

The long axes of the two articular surfaces (shown by dotted outlines here) and the sinus tarsi are directed anteromedially.

The subtalar joint allows limited ROM in two planes, as shown:

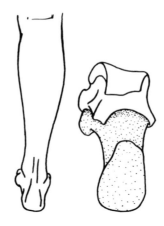

• neutral (anatomical) position

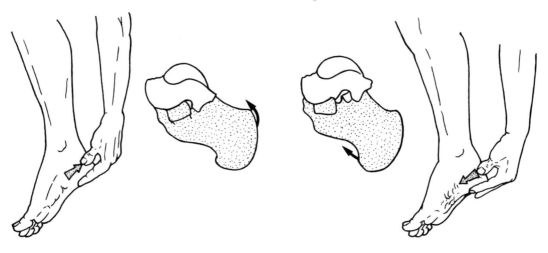

• plantar flexion

• dorsiflexion

RIGHT ANKLE, POSTERIOR VIEW:

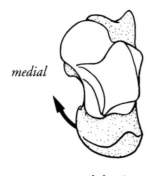

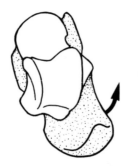

medial

lateral

• abduction

(heel moves medially, toes
move laterally; see page 256)

• adduction

(heel moves laterally,
toes move medially)

Actually, because of the structure of the articulating surfaces, these movements tend to be combined around an imaginary line called the axis of Henke (a German anatomist). This axis enters the posterolateral tuberosity of calcaneus, runs anterosuperomedially, and exits through the medial neck of talus.

axis of Henke

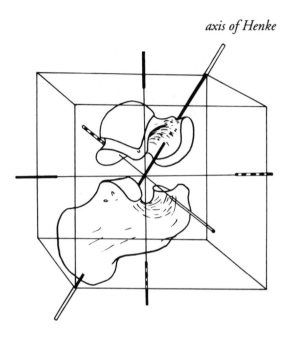

The movement called **inversion** occurs around this axis, and is a combination of adduction and plantar flexion. The opposite movement, **eversion**, is a combination of abduction and dorsiflexion.

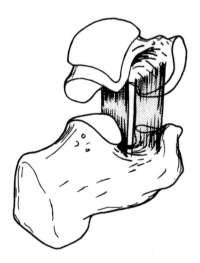

The capsule of the subtalar joint wraps around the articulating surfaces, and is continous anteriorly with the capsule of the transverse tarsal joint. In addition to the interosseous talocalcaneal ligament, there are weaker posterior and anterior ligaments between the two bones.

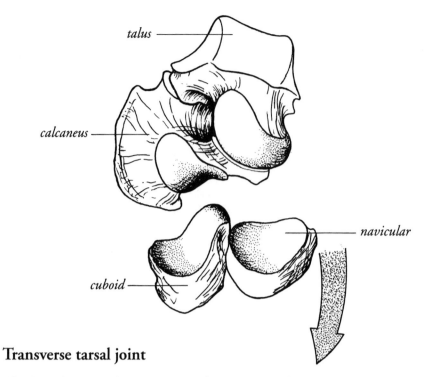

talus

calcaneus

navicular

cuboid

Transverse tarsal joint

This joint between the posterior and anterior tarsals comprises the medial **talocalconeonavicular** and lateral **calcaneocuboid joints**. The medial part is higher and involves a convex surface of the talar head fitting against a concave surface of navicular. The lateral part is lower and involves an S-shaped surface of calcaneus (concave medially, convex laterally) fitting against a corresponding surface of cuboid (above).

Seen from above, the transverse tarsal joint is curved rather than straight (right).

The basic movements here are inversion and eversion. Adjustments of the foot when walking on uneven ground depend mainly on movements of the transverse tarsal and more distal joints.

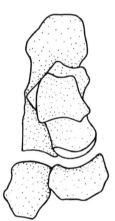

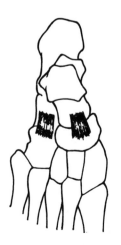

Superiorly, the joint is reinforced by **talonavicular** and **calcaneocuboidal ligaments** (right).

Laterally, the **bifurcate ligament** runs from calcaneus and spreads out vertically on navicular and horizontally on cuboid (the surfaces of these two bones are roughly perpendicular at this junction).

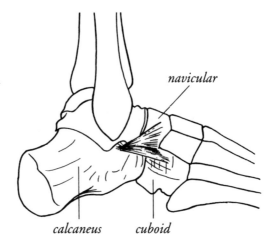

navicular

calcaneus   *cuboid*

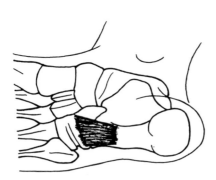

Inferiorly, the transverse tarsal joint is reinforced by three ligaments. The **plantar calcaneocuboid (short plantar) ligament** runs from calcaneus to the proximal cuboid.

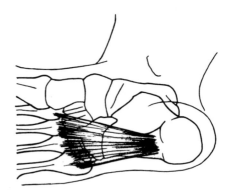

The **long plantar ligament** (located superficial to the short plantar ligament) runs from calcaneus to a ridge on cuboid, passes over the groove (thereby forming a groove for the peroneus longus tendon), and continues on to attach to the bases of metatarsals II-V.

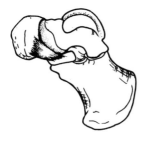

The long plantar ligament is quite strong and helps support the arches of the foot. Inferomedially, the **plantar calcaneonavicular ligament** runs from the sustentaculum tali to navicular, and helps support the talar head.

Cuneiform joints

The cuneiforms articulate with cuboid, navicular, and each other. These joints all share a common synovial cavity. There are dorsal and plantar ligaments at each junction, plus interosseous ligaments between individual cuneiforms, and between cuboid and the lateral cuneiform.

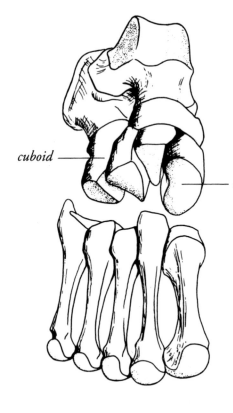

cuboid

cuneiforms

Tarsometatarsal joints

Here, the cuneiforms and cuboid meet the bases of the metatarsals to form gliding joints. Collectively, the junction is irregular, not straight.

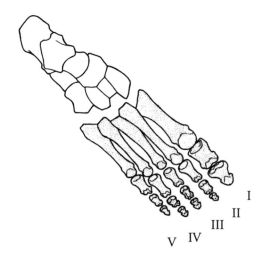

I

II

III

V IV

Metatarsal I articulates only with the medial cuneiform, where it has its own synovial cavity (the others share a single cavity). Metatarsal II fits into a notch formed by all three cuneiforms. Metatarsal III meets the lateral cuneiform, and metatarsals IV and V fit against cuboid (above). Plantar and dorsiflexion are the possible movements at these joints. ROM is in the increasing order II, III, I, IV, V.

The joints are reinforced by numerous dorsal (shown at left) and plantar (not shown) ligaments.

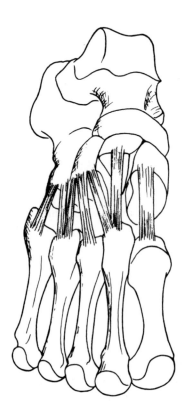

Metatarsophalangeal and interphalangeal joints

The **metatarsophalangeal joints** allow three types of movement:

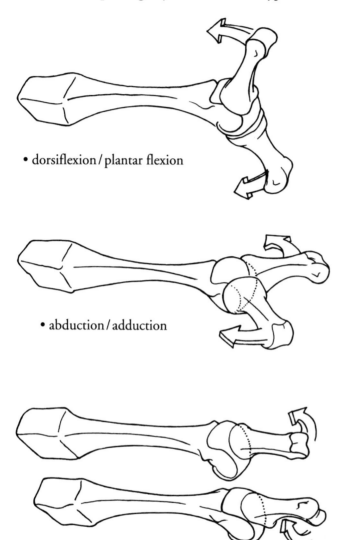

• dorsiflexion / plantar flexion

• abduction / adduction

• medial / lateral rotation
(these are mainly passive)

Dorsiflexion has greater ROM than plantar flexion, and the associated muscles are stronger. For standing on tiptoe, walking, etc., strong dorsiflexion is needed.

The proximal **interphalangeal joint** is a hinge. It allows plantar flexion but not dorsiflexion.

The distal interphalangeal joint is also a hinge, but allows both plantar and dorsiflexion.

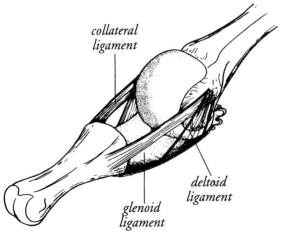

collateral ligament

deltoid ligament

glenoid ligament

Ligaments of the metatarsophalangeal and interphalangeal joints have the same general plan:

• two collateral ligaments, inserting on proximal tubercles of the distal bone;

• a plantar "glenoid" ligament, which folds onto itself during plantar flexion;

• a fan-shaped "deltoid" ligament, running from the tubercle to the glenoid ligament.

In the big toe, the metatarsal and phalanges are larger than in toes II-V, and there are two phalanges instead of three. This toe plays an important role in walking or running, especially in the digitigrade phase (i.e., when the toes are in contact with the ground). The disproportionate size of metatarsal I can lead to instability or medial pain during tiptoeing or prolonged walking.

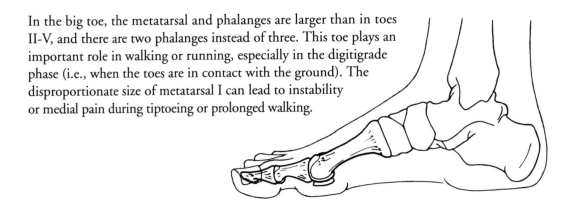

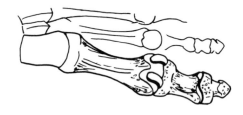

Two small sesamoid bones are located in the plantar cartilage on the head of metatarsal I. They act as shock-absorbers during weight-bearing.

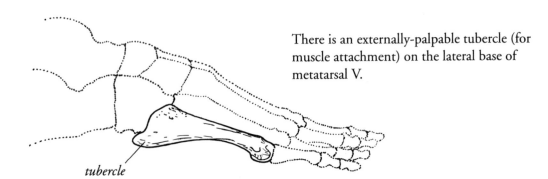

tubercle

There is an externally-palpable tubercle (for muscle attachment) on the lateral base of metatarsal V.

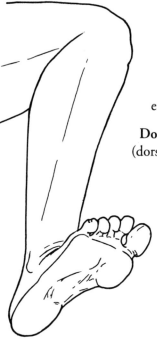

Movements

We can speak of the following movements in reference to the entire foot, or to some specific region or joint.

Dorsiflexion is a decrease in the angle between the superior (dorsal) surface of the foot and the anterior leg.

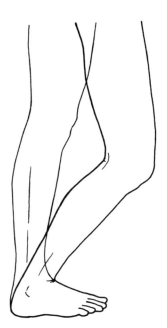

ROM for dorsiflexion is greater when the knee is flexed, and less when the knee is extended. Why? Because there is greater tension in the gastrocnemius when the knee is extended (see page 265).

Plantar flexion, also called extension, is an increase in the angle between the dorsal surface of the foot and the anterior leg. Dorsiflexion and plantar flexion occur mainly at the ankle joint.

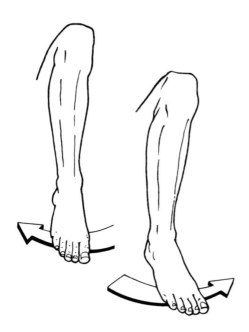

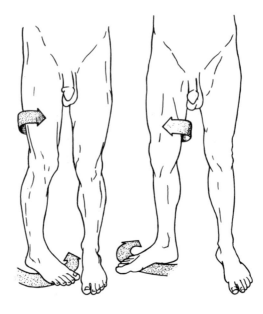

In **abduction** and **adduction**, the distal end of the foot moves away from and toward the median plane, respectively.

These movements can be amplified by or confused with medial and lateral rotation of the hip (when the knee is extended)...

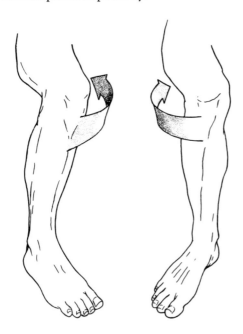

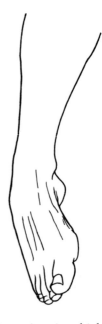

...or rotation of the flexed knee. In either of these situations, you will notice movement of the tibial tuberosity. Remember that the ankle joint is a pure hinge, capable only of dorsiflexion/plantar flexion. Therefore, abduction/adduction occur at the subtalar and more distal joints.

Eversion, in which the sole of the foot is directed away from the median plane, is a combination of abduction and dorsiflexion.

Inversion, in which the sole is directed toward the median plane, is a combination of adduction and plantar flexion.

Muscles

The **extrinsic muscles** of the foot originate on the femur, tibia, or fibula (left), and insert on foot bones via long tendons. They are all polyarticular. The **intrinsic muscles** are short muscles which run between nearby foot bones. They are mostly on the plantar side of the bones, and comprise the fleshy mass of the sole.

The ankle has an **extensor retinaculum** and **flexor retinaculum**, homologous to the same-named structures of the wrist, to help keep the tendons of certain extrinsic muscles in place. The extensor retinaculum consists of a superior band running from fibula to tibia anteriorly (just above the lateral and medial malleoli), and a Y-shaped inferior band running from calcaneus to the medial malleolus and the medial edge of the plantar aponeurosis (right).

tibialis anterior

extensor digitorum longus

extensor hallucis longus

extensor retinaculum

LATERAL VIEW:

ANTERIOR VIEW:

tendon of peroneus tertius

peroneus longus

MEDIAL VIEW:

flexor digitorum longus

tibialis posterior

tibialis anterior

tendon of flexor hallucis longus

peroneus brevis

flexor retinaculum

lateral malleolus

peroneal retinacula

The flexor retinaculum runs from the medial malleolus to calcaneus (above). There is also a **peroneal retinaculum** for the peroneus muscle tendons (left). Its superior band runs from the lateral malleolus to calcaneus, and the inferior band between two spots on the anterolateral calcaneus.

Extrinsic anterior muscles

Tibialis anterior originates from the lateral condyle and supero-lateral shaft of tibia, passes under the extensor retinaculum, and inserts on the medial cuneiform (inferomedial surface) and base of metatarsal I. This muscle is the strongest dorsiflexor.

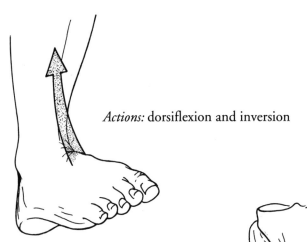

Actions: dorsiflexion and inversion

Extensor hallucis longus arises from the central medial fibula and interosseous membrane, passes under the extensor retinaculum, and inserts dorsally on distal phalanx I.

Action:
dorsiflexion of big toe and foot

Extensor digitorum longus originates from the lateral tibial condyle, most of the anterior fibular shaft, and interosseous membrane. Its tendon passes under the extensor retinaculum, splits into four parts, and inserts on toes II-V. Each of the four tendons further splits into two slips attaching to the sides of the middle phalanx, and a central slip attaching to the base of the distal phalanx. This is reminiscent of the extensor digitorum of the hand (see page 167).

Action: dorsiflexion of toes II-V and foot

Action of this muscle is completed by various intrinsic muscles (described later) which attach to its insertions.

Peroneus tertius is an insignificant muscle, absent in some individuals. It is essentially a part of extensor digitorum longus which fails to reach the toes. It arises from the anteroinferior fibula and inserts on metatarsal V.

Actions: dorsiflexion and eversion

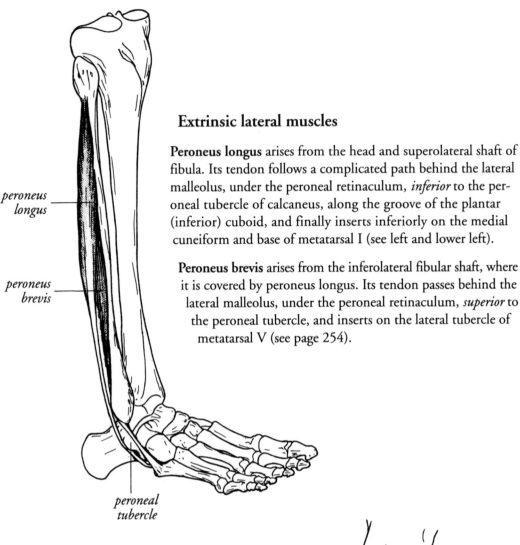

peroneus longus

peroneus brevis

peroneal tubercle

Extrinsic lateral muscles

Peroneus longus arises from the head and superolateral shaft of fibula. Its tendon follows a complicated path behind the lateral malleolus, under the peroneal retinaculum, *inferior* to the peroneal tubercle of calcaneus, along the groove of the plantar (inferior) cuboid, and finally inserts inferiorly on the medial cuneiform and base of metatarsal I (see left and lower left).

Peroneus brevis arises from the inferolateral fibular shaft, where it is covered by peroneus longus. Its tendon passes behind the lateral malleolus, under the peroneal retinaculum, *superior* to the peroneal tubercle, and inserts on the lateral tubercle of metatarsal V (see page 254).

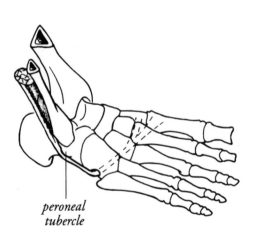

peroneal tubercle

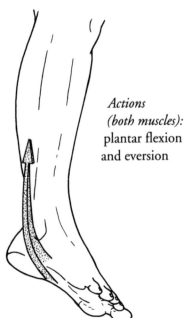

Actions (both muscles): plantar flexion and eversion

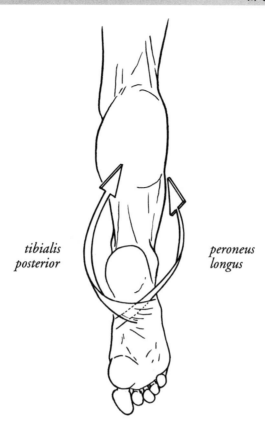

*tibialis
posterior*

*peroneus
longus*

The tendons of peroneus longus and tibialis posterior (see next page), coming from opposite sides, form a "sling" under the middle part of the foot which is crucial in supporting the arches (above). Peroneus longus and brevis both strengthen and support the lateral arch (the weight-bearing arch; see page 243), stabilizing the ankle and preventing loss of balance laterally when standing, especially when raised on tiptoe (below).

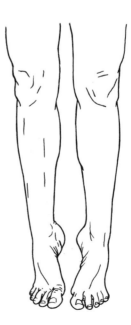

Extrinsic posterior muscles

Flexor digitorum longus originates from the postero-medial tibial shaft, runs posterior to the medial malleolus and sustentaculum tali, along the plantar surface of the foot, and inserts on distal phalanges II-V.

Tibialis posterior is the deepest calf muscle. It arises from the postero-superior tibial and fibular shafts and interosseous membrane, and passes posterior to the medial malleolus and anterior to sustentaculum tali. Its primary insertion is on a prominent medial tubercle of navicular (see page 241), but it also inserts on cuboid, lateral cuneiform, and metatarsals II-IV.

Actions: plantar flexion of toes II-V and ankle, inversion of foot, support of arches. This is the most powerful flexor of toes II-V.

Actions: plantar flexion, inversion, support of arches. The role of tibialis posterior, in conjunction with peroneus longus, in forming a "sling" for the middle foot, was mentioned on the preceding page.

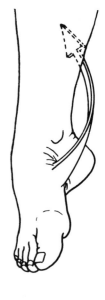

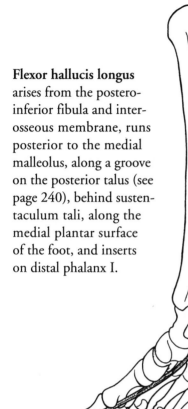

Flexor hallucis longus arises from the postero-inferior fibula and interosseous membrane, runs posterior to the medial malleolus, along a groove on the posterior talus (see page 240), behind sustentaculum tali, along the medial plantar surface of the foot, and inserts on distal phalanx I.

This muscle is important in the propulsion phase of walking (see page 243), and also in preventing anterior loss of balance when standing on tiptoe.

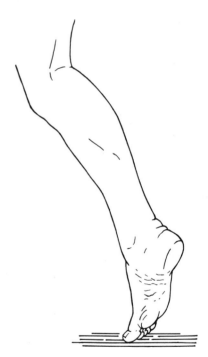

Actions: plantar flexion of big toe and ankle, inversion, support of medial arch

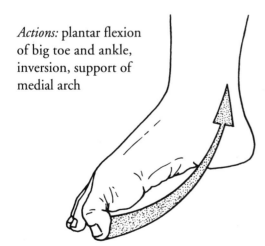

Gastrocnemius and soleus comprise the superficial, bulky group of posterior calf muscles known collectively as the **triceps surae**. They insert on calcaneus via the Achilles tendon and are the primary plantar flexors of the ankle. Paradoxically, the Achilles tendon is both the strongest and the most frequently-ruptured tendon of the human body.

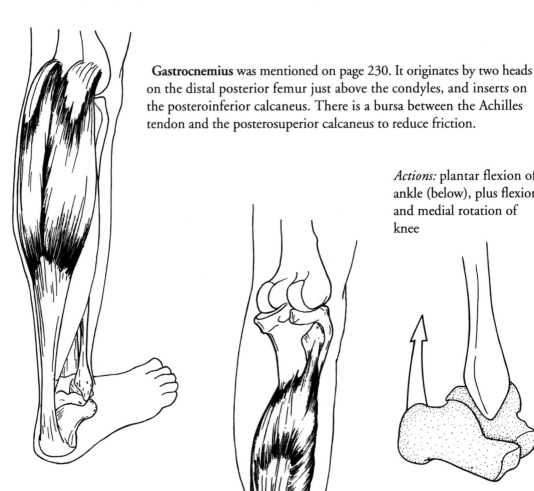

Gastrocnemius was mentioned on page 230. It originates by two heads on the distal posterior femur just above the condyles, and inserts on the posteroinferior calcaneus. There is a bursa between the Achilles tendon and the posterosuperior calcaneus to reduce friction.

Actions: plantar flexion of ankle (below), plus flexion and medial rotation of knee

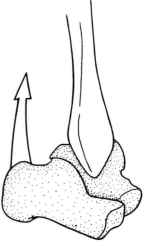

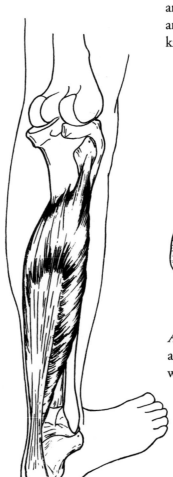

Soleus arises broadly from the posterosuperior tibia and fibula and merges with gastrocnemius to insert on calcaneus via the Achilles tendon.

Action: plantar flexion of ankle, especially during walking

The plantar flexion produced by the triceps surae tends to be associated with inversion and adduction of the foot. Why? Because of the shape of the articulating surfaces of the subtalar joint, and the "axis of Henke" (see page 249).

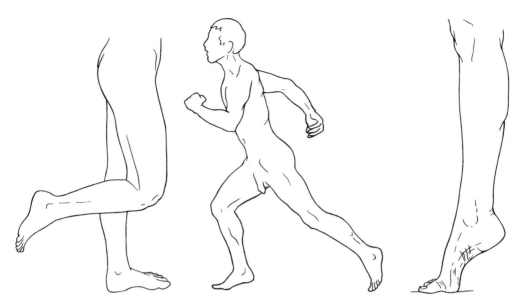

Since gastrocnemius crosses the knee, the position of the knee affects its efficiency as a plantar flexor of the ankle. When the knee is very flexed, the gastrocnemius is slack and therefore inefficient as an ankle flexor.

When the knee is extended or only slightly flexed (the position taken by the "propelling leg" at the start of a race), the gastrocnemius is more taut and more efficient as an ankle flexor.

For standing on tiptoe, plantar flexion of the ankle by the triceps surae is necessary, but not sufficient by itself. The Achilles tendon affects only calcaneus. Obviously, many other muscles inserting on the more distal bones are also involved in this movement.

Dorsiflexion at the ankle stretches the soleus.

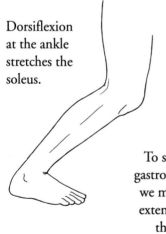

To stretch the gastrocnemius, we must add extension of the knee.

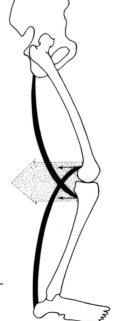

Interestingly, when the knee is flexed and the leg and foot are bearing the weight of the body, the gastrocnemius and hamstrings combine their forces to become *extensors* of the knee, i.e., returning it to anatomical position. When the foot is not bearing the body's weight, these muscles act as *flexors* of the knee (right).

Intrinsic muscles

Extensor digitorum brevis is the only dorsal intrinsic muscle. It arises from the anterosuperolateral calcaneus and divides into four bodies. The medial tendon inserts on proximal phalanx I, while the other three merge laterally with the tendons of extensor digitorum longus inserting on toes II-IV.

Action: dorsiflexion of toes I-IV, in cooperation with extensor digitorum longus

There are several layers of plantar intrinsic muscles.

The **interossei** form the fourth (deepest) layer, and occupy the spaces between the metatarsals, where they originate. There are four dorsal interossei (i.e., arising closer to the dorsal surface of the foot)...

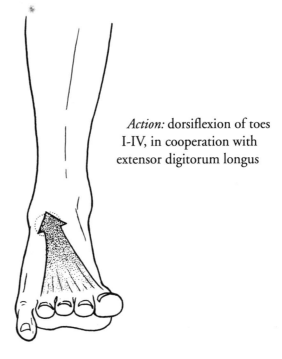

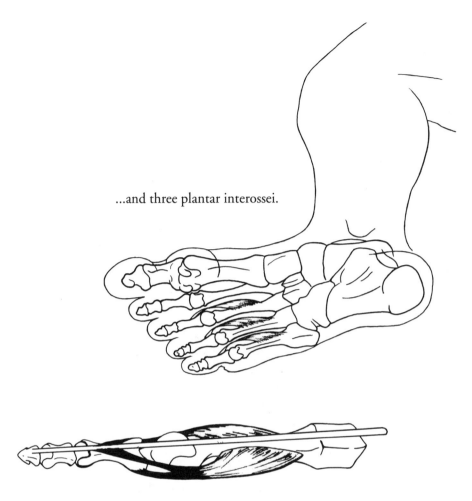

...and three plantar interossei.

They insert on the proximal phalanges at the base (plantar side) and at the extensor digitorum longus tendon (dorsal side).

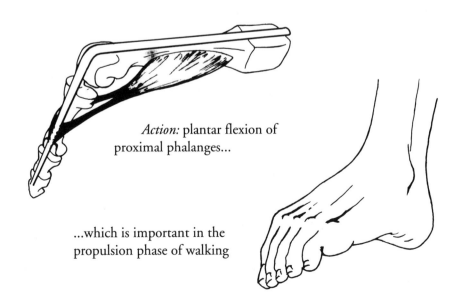

Action: plantar flexion of proximal phalanges...

...which is important in the propulsion phase of walking

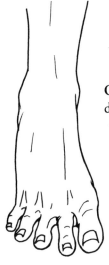

By contracting on one side or the other, the interossei also help spread the toes apart or pull them back together. Their origins (proximal attachments) help prevent the metatarsals from spreading apart, and maintain the transverse arch of the foot.

Quadratus plantae (also called flexor digitorum accessorius) arises from the body of calcaneus, and inserts on the posterolateral border of the flexor digitorum longus tendon near its division into four parts (right). It belongs to the second layer.

Action: redirects the pull of the flexor digitorum longus tendons to be more in line with the axes of the toes

quadratus
plantae

lumbricals

The **lumbricals** are four small muscles (in the second layer) running from the flexor digitorum longus tendons to the dorsal parts of the extensor digitorum longus tendons. They work with the interossei to plantar flex the interphalangeal joints when the metatarsophalangeal joints are held in flexion, i.e., push the toes off the ground during the propulsion phase.

Flexor digitorum brevis is part of the first (most superficial) layer. It arises from the posteroinferior tuberosity of calcaneus, splits into four parts, and inserts laterally on middle phalanges II-V. Each tendon is "perforated" to allow passage of the flexor digitorum longus tendons to the distal phalanges.

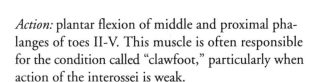

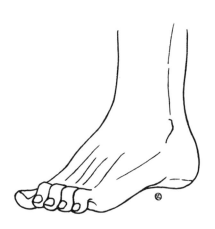

Action: plantar flexion of middle and proximal phalanges of toes II-V. This muscle is often responsible for the condition called "clawfoot," particularly when action of the interossei is weak.

Flexor hallucis brevis (third layer) originates from cuboid and the two lateral cuneiforms and inserts via two tendons on either side of proximal phalanx I.

Action: plantar flexion of proximal phalanx of big toe

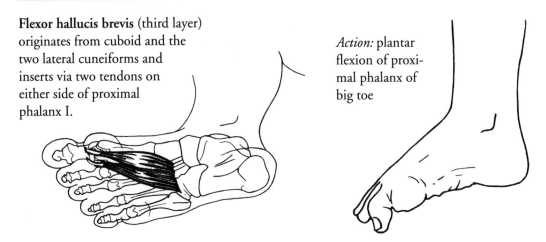

Abductor hallucis (first layer) arises from calcaneus and the medial flexor retinaculum, and inserts medially on proximal phalanx I. [Note: when we speak of abduction or adduction of toes, the reference is the axis of the second toe, not the median plane of the body.] This powerful muscle plays an important role in keeping the big toe properly aligned during walking. It opposes "hallux valgus."

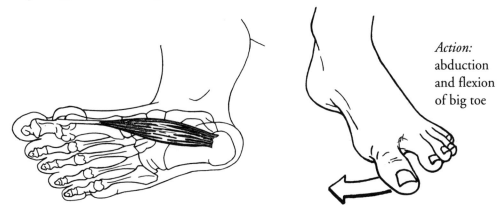

Action: abduction and flexion of big toe

Adductor hallucis (third layer) has two heads. The oblique head arises from the bases of metatarsals II-IV, and the transverse head from the capsules of metatarsophalangeal joints III-V. The tendon merges with the lateral tendon of flexor hallucis brevis and inserts laterally on the base of proximal phalanx I. This is the muscle primarily responsible for "hallux valgus," an abnormal condition in which metatarsal I is permanently abducted and proximal phalanx I adducted, such that the big toe overlaps the second toe.

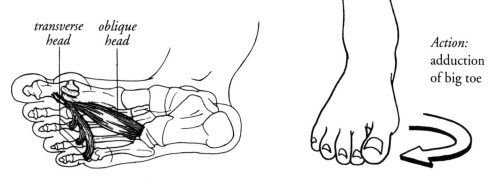

transverse head   oblique head

Action: adduction of big toe

Flexor digiti minimi brevis (third layer) arises from the base of metatarsal V and inserts on the base of proximal phalanx V.

Action: plantar flexion of little toe

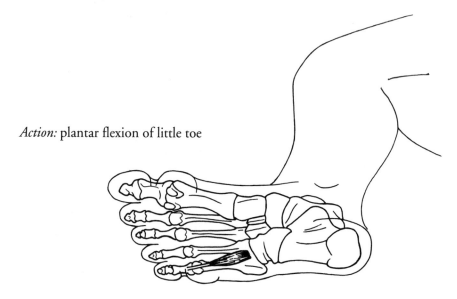

Abductor digiti minimi (first layer) originates from the posteroinferior calcaneus and inserts laterally on the base of proximal phalanx V.

Action: abduction of little toe, support of lateral arch

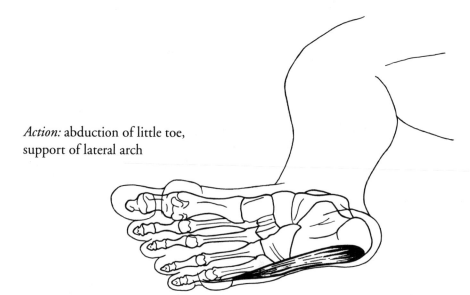

Summary of movements/stability

As in Chapter 6, it will be helpful at this point to summarize the muscles involved in specific movements of the foot. The arrows represent the forces produced by the various muscles.

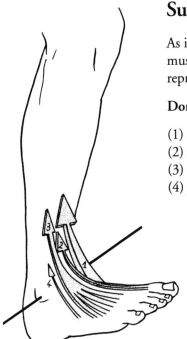

Dorsiflexion:

(1) tibialis anterior
(2) extensor hallucis longus
(3) extensor digitorum longus
(4) peroneus tertius

Plantar flexion:

(1) peroneus longus
(2) peroneus brevis
(3) triceps surae
(4) flexor hallucis longus
(5) tibialis posterior
(6) flexor digitorum longus

Inversion/adduction:

(1) extensor hallucis longus
(2) tibialis anterior
(3) tibialis posterior
(4) flexor digitorum longus
(5) flexor hallucis longus
not shown:
    triceps surae

Eversion/abduction:

(1) peroneus longus and brevis
(2) peroneus tertius
(3) extensor digitorum
        longus (lateral part)

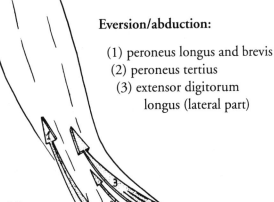

Notice that opposing actions are not "balanced." Plantar flexion is dominant over dorsiflexion, and inversion/adduction is dominant over eversion/abduction.

Stability of ankle joint

As explained on page 245, the talus fits more snugly into the "pincer" (formed by the distal tibia and fibula) during dorsiflexion than plantar flexion.

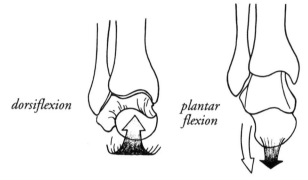

dorsiflexion   *plantar flexion*

To compensate for the loss of stability during plantar flexion, the shape of the "pincer" is modified by four muscles which tend to pull down the fibula: peroneus longus, peroneus brevis, extensor hallucis longus, and tibialis posterior (right). The fit of the pincer on the plantar flexed talus is improved when the lateral malleolus is lowered (below).

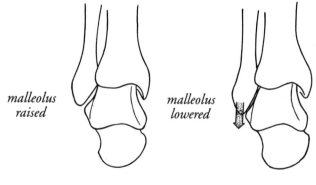

malleolus raised   *malleolus lowered*

A related factor is the "tightening" of the two sides of the "pincer," produced in part by extensor hallucis longus and tibialis posterior (left).

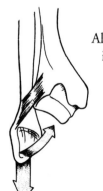

Also, when the fibula is lowered, the distal tibiofibular ligaments come under tension, which automatically tightens the "pincer."

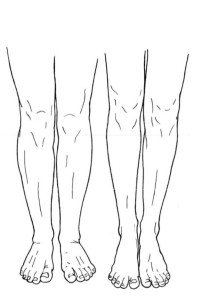

These effects on the "pincer" by muscles and ligaments stabilize the ankle during plantar flexion, e.g., standing on tiptoe.

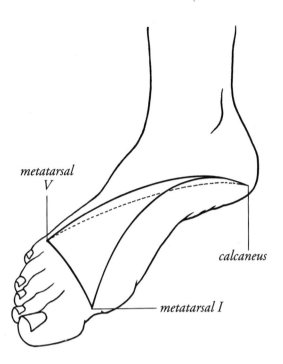

Arches of foot

The foot does not rest flat on the ground. The structural arrangement of bones, ligaments, and tendons in the foot results in three arches. These arches are crucial in giving the foot flexibility, absorbing shock, distributing the weight of the body, and adapting the shape of the plantar foot to the surfaces it encounters during walking, climbing, etc.

When standing, the weight of the body is essentially distributed among three points. The posteroinferior tuberosity of calcaneus receives most of the weight. The secondary weight-bearing point is the head of metatarsal I, a relatively massive bone. The third point, the head of metatarsal V, supports the least weight.

The **medial arch** is formed primarily by five bones (calcaneus, talus, navicular, medial cuneiform, metatarsal I), four ligaments (talocalcaneal, calcaneonavicular, and small ligaments joining the cuneiform to navicular and metatarsal), and four muscles (abductor hallucis, tibialis posterior, peroneus longus, flexor hallucis longus).

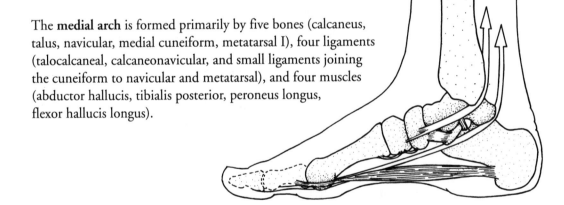

Flexor hallucis longus has three functions with respect to the medial arch:

- stretches the arch like the string of a bow
- supports calcaneus by passing under the sustentaculum tali
- supports talus by passing along its posterior groove.

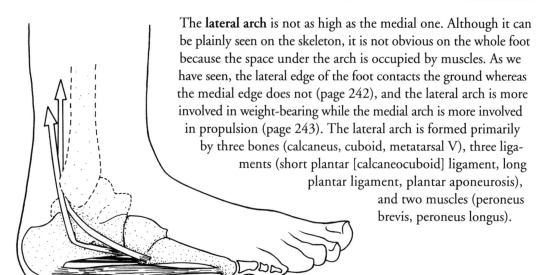

The **lateral arch** is not as high as the medial one. Although it can be plainly seen on the skeleton, it is not obvious on the whole foot because the space under the arch is occupied by muscles. As we have seen, the lateral edge of the foot contacts the ground whereas the medial edge does not (page 242), and the lateral arch is more involved in weight-bearing while the medial arch is more involved in propulsion (page 243). The lateral arch is formed primarily by three bones (calcaneus, cuboid, metatarsal V), three ligaments (short plantar [calcaneocuboid] ligament, long plantar ligament, plantar aponeurosis), and two muscles (peroneus brevis, peroneus longus).

Peroneus longus has two functions here:

• supports calcaneus by passing under peroneal tubercle

• supports cuboid.

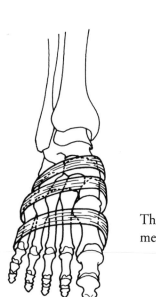

The **transverse arch** (left) is most visible around the middle of the metatarsals. It is symbolized in this picture by straps.

As you would expect, it is higher on the medial (navicular) side than the lateral (cuboid) side. Its muscular support comes primarily from adductor hallucis (transverse head), peroneus longus, tibialis posterior, and the interossei.

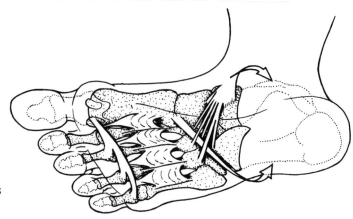

Walking

As in the previous chapter, let us examine the sequential actions of walking, focusing now on the foot muscles.

foot hits the ground heel-first, then rolls forward

contraction of dorsiflexors during rolling movement

body's weight on foot; foot in full contact with ground

contraction of muscles supporting the three arches

heel leaves ground; propulsion by distal foot

contraction of triceps surae and other plantar flexors; contraction of intrinsic plantar muscles

toes leave ground (big toe last)

contraction of flexor digitorum longus, then flexor hallucis longus

entire foot is off the ground; foot moves rapidly forward

brief moment of muscular relaxation while foot is off ground; contraction of dorsiflexors to prevent toes from touching ground during forward movement of foot

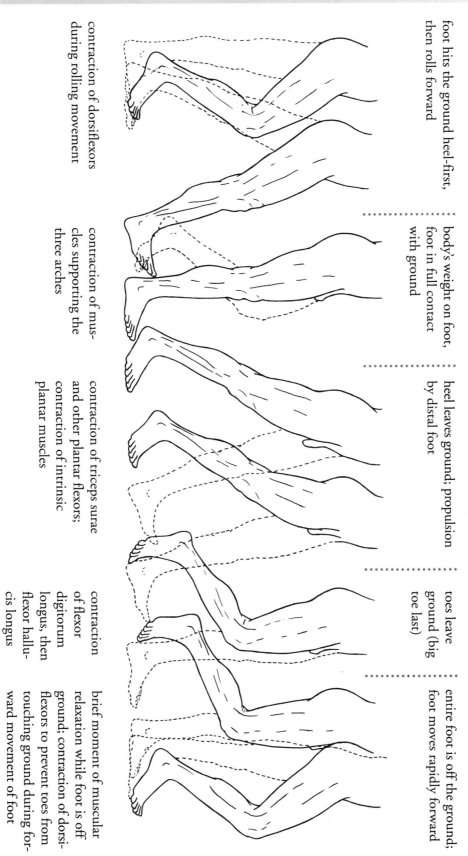

Index

G

H